SIMPLE & SATISFYING

VEGETABLES

A GUIDE TO BUYING, STORING, AND COOKING VEGETABLES

First published in the United States of America in 2023 by Christopher J. Riesbeck

ISBN: 979-8-9882293-7-7
Library of Congress Control Number: 2023907252
1st edition

Contents

Foreword

Welcome to *Simple and Satisfying - Vegetables:* A Guide to Buying, Storing, and Cooking Vegetables. A cookbook designed to provide a basic introduction to all the commonly found vegetables. Each recipe is designed to be as simple as possible to appreciate the simplicity and elegance of these vegetables. Each section of the cookbook focuses on the essentials of a specific vegetable. Learning the fundamentals of a vegetable from basic comprehensive information will strengthen your foundation of culinary knowledge.

Each vegetable section presents three enticing recipes that demonstrate the adaptability and palatability of the highlighted vegetable. Each recipe employs a distinct, favored cooking technique while utilizing minimal ingredients. Additionally, the final cooking method for each vegetable involves an air fryer!

Whether you are an experienced vegetable enthusiast or entirely new, this book serves as an ideal companion. Remember each recipe is designed to be basic so if you want to try something more complex, listen to what your palate tells you to do so let us embark on this journey and explore the delectable world of vegetables together!

-A.H. Uman

CONVERSIONS

Standard Unit	Metric Unit
1 teaspoon	5 ml
1 tablespoon	15 ml
1 fluid ounce	30 ml
1 cup	240 ml
1 pint	473 ml
1 quart	946 ml
1 gallon	3.785 L
1 ounce	28.35 g
1 pound	453.59 g

Introduction

Dear Reader,

It is with immense joy and gratitude that I present to you *Simple and Satisfying: Vegetables*, a labor of love over a decade in the making. This journey began with a humble seed of an idea, sprouted from my lifelong passion for vegetables and my desire to share their beauty and versatility with others. Over the years, this seed has grown into a bountiful garden, nourished by dedication, inspiration, and countless hours spent perfecting this book.

This cookbook is more than just a collection of recipes; it is a testament to the transformative power of vegetables and the incredible impact they can have on our lives. Each page has been lovingly crafted to provide you with not only the knowledge and techniques to create delicious and nutritious meals, but also a sense of connection to the earth and its bounty.

As you embark on this culinary adventure, I invite you to immerse yourself in the vibrant world of vegetables. Savor the rich colors, the diverse textures, and the tantalizing flavors that await you in every recipe. Allow yourself to be inspired by the simple elegance of these dishes and the endless possibilities they present.

I have poured my heart and soul into this book, and I hope that it will serve as a cherished companion on your journey towards healthier, more sustainable eating. It is my sincerest wish that you experience the freshest of vegetables and the warmest of company to share them with, creating unforgettable memories and fostering a deep appreciation for the world around us.

With love and gratitude,

A.H. Uman

SEASONAL VEGETABLE GUIDE

THE BEST TIME TO ENJOY EACH VEGETABLE IS AT ITS PEAK OF FRESHNESS AND FLAVOR. THE CHART BELOW SHOWCASES WHICH VEGETABLES YOU'LL FIND DURING EACH SEASON, MAKING IT EASY FOR YOU TO PLAN MEALS AND SHOP ACCORDINGLY. FROM SPRING'S VIBRANT GREENS TO WINTER'S HEARTY ROOTS, BEGIN LEARNING ABOUT THE BOUNTY OF EACH SEASON WITH THIS HANDY, VISUAL GUIDE.

SPRING

- Artichoke
- Arugula
- Asparagus
- Beetroot
- Bok Choy
- Carrots
- Green Beans
- Kale
- Leeks
- Lettuce (Romaine)
- Peas
- Radish
- Spinach
- Swiss Chard
- Turnip
- Watercress

SUMMER

- Bell Pepper
- Broccoli
- Cabbage
- Cauliflower
- Celery
- Chili Pepper
- Collard Greens
- Corn
- Cucumber
- Eggplant
- Garlic
- Ginger
- Jicama
- Kohlrabi
- Mushroom
- Okra
- Onion
- Parsnip
- Potato
- Pumpkin
- Shallot
- Squash
- Sunchoke
- Sweet Potato
- Tomato
- Zucchini

FALL

- Beetroot
- Brussels Sprouts
- Celery
- Collard Greens
- Garlic
- Leeks
- Parsnip
- Radicchio
- Rutabaga
- Spinach
- Swiss Chard
- Turnip
- Watercress

WINTER

- Beetroot
- Brussels Sprouts
- Cabbage
- Collard Greens
- Kale
- Leeks
- Parsnip
- Spinach
- Swiss Chard
- Turnip
- Watercress

ARTICHOKE

History Artichokes are a fascinating vegetable with a rich history, dating back to ancient Greece and Rome. They have long been enjoyed for their unique, earthy taste and delicate texture. Cultivated in the Mediterranean region for centuries, artichokes made their way to the United States in the 19th century, where they are now primarily grown in California.

Can you eat it raw? While it is possible to eat artichokes raw, they are generally tough and fibrous when uncooked. Cooking helps to soften the artichoke and bring out its complex flavors.

Growing and harvesting Artichokes thrive in mild climates with cool, foggy summers and temperate winters. They grow best in well draining, fertile soil with consistent moisture. The artichoke plant produces large, spiky leaves, and the edible part, called the bud, is harvested before the flower fully blooms. Harvesting typically occurs when the bud is 3 to 5 inches in diameter.

Ripeness An artichoke is ripe when its bud is tightly closed, and the leaves are green and firm to the touch. When the leaves begin to spread and turn purple or brown, the artichoke is past its prime and may become tough and bitter.

Spoilage A spoiled artichoke may have a slimy texture or exhibit mold growth, often accompanied by an unpleasant odor. Discolored or soft leaves can also be an indicator of spoilage.

Storing To store fresh artichokes, place them in a plastic bag with a few drops of water to maintain humidity, and refrigerate for up to one week.

Preserving The simplest way to preserve artichokes is to blanch and freeze them. First, trim and clean the artichokes, then blanch them in boiling water for 3-5 minutes. Quickly cool them in an ice bath, drain, and store in airtight freezer containers for up to six months.

Steamed Artichoke

Serves: 2
Prep Time: 10 minutes
Cook Time: 40 minutes

1 large artichoke
15 ml extra virgin olive oil
Fresh lemon juice (optional)
120 ml water
Salt and pepper, to taste

Directions

1. Cut the top 2.5 cm off each artichoke and trim the stem. Remove the small leaves from the base.
2. Fill a pot with a steamer basket with 2.5 cm of water and bring to a boil over high heat.
3. Place the artichokes in the steamer basket, cover, and steam for 30-40 minutes, or until the leaves are tender and can be easily pulled off.
4. In a small bowl, whisk together the olive oil, lemon juice, salt and pepper.
5. Serve the artichokes hot with the olive oil mixture for dipping.

Notes

For a tangy variation, minced garlic and Dijon mustard to the olive oil mixture.

Roasted Artichoke

Serves: 2
Prep Time: 10 minutes
Cook Time: 30 minutes

1 large artichokes
30 ml extra virgin olive oil
Balsamic vinegar (optional)
Salt and pepper, to taste

Directions

1. Preheat the oven to 200°C (400°F).
2. Line a baking sheet with parchment paper.
3. Cut the top 2.5 cm off each artichoke and trim the stem. Remove the small leaves from the base. Cut each artichoke in half lengthwise and use a spoon to scoop out the fuzzy choke in the center.
4. In a small bowl, whisk together the olive oil, balsamic vinegar, salt and pepper.
5. Place the artichokes cut side up on the prepared baking sheet and brush with the olive oil mixture.
6. Roast for 25-30 minutes or until the edges are browned and the artichokes are tender.

Notes

For a more savory variation, add minced garli, dried oregano, and red pepper flakes to the olive oil mixture.

Air Fried Artichokes

Serves: 2
Prep Time: 10 minutes
Cook Time: 15 minutes

1 large artichokes
30 ml extra virgin olive oil
Apple cider vinegar (optional)
Salt and pepper, to taste

Directions

1. Preheat the air fryer to 200°C (400°F).
2. Cut the top 2.5 cm off each artichoke and trim the stem. Remove the small leaves from the base.
3. Cut each artichoke in half lengthwise and use a spoon to scoop out the fuzzy choke in the center.
4. In a small bowl, whisk together the olive oil, vinegar, salt and pepper.
5. Place the artichokes in the air fryer basket on aluminum foil, cut side up, and brush with the olive oil mixture.
6. Air fry for 12-15 minutes or until the edges are crispy and the artichokes are tender.

Notes

For a zesty variation, add minced garlic clove, grated lemon zest, and smoked paprika to the olive oil mixture.

ARUGULA

History Arugula, also known as rocket or roquette, is a leafy green vegetable with a peppery flavor and a slightly nutty undertone. It is native to the Mediterranean region and has been enjoyed for centuries, dating back to ancient Rome, where it was believed to possess aphrodisiac properties.

Can you eat it raw? Arugula can be consumed raw and is often used in salads, mixed with other greens, or as a garnish for various dishes. Its bold and peppery flavor adds a unique and refreshing taste to any meal.

Growing and harvesting Arugula is a cool season crop that thrives in well draining, fertile soil with a pH of 6.0-7.0. It prefers full sun but can tolerate light shade. The seeds are sown directly into the ground or into starter pots and are usually harvested within 4-6 weeks after planting. The leaves are picked when they are young and tender, as they can become more bitter and tough as they mature.

Ripeness To determine the ripeness of arugula, look for vibrant green leaves that are firm and free of yellowing or wilting. The leaves should be crisp and tender to the touch.

Spoilage Arugula spoils when the leaves become slimy, excessively wilted, or develop a strong, unpleasant odor. It is important to discard spoiled arugula to avoid foodborne illnesses.

Storing To store arugula, wrap the unwashed leaves in a paper towel and place them in a plastic bag or airtight container. Store in the refrigerator for up to 5 days. For longer storage, consider freezing or preserving the arugula.

Preserving To preserve arugula, blanch the leaves in boiling water for 30 seconds, then immediately transfer them to an ice bath to stop the cooking process. Drain the leaves, pat them dry, and store them in a freezer safe bag or container. Freeze for up to 6 months.

ARUGULA SALAD

Serves: 2
Prep Time: 10 minutes
Cook Time: 0

240 g arugula, washed and dried
60 ml extra virgin olive oil
30 ml fresh lemon juice
Salt and pepper, to taste

Directions

1. In a large salad bowl, whisk together the olive oil, lemon juice, salt and pepper to make the vinaigrette.
2. Add the arugula leaves to the bowl and toss to coat them evenly with the vinaigrette.
3. Serve immediately.

Notes

Add thinly sliced red onions, cherry tomatoes, or shaved Parmesan cheese for extra flavor and texture.

SAUTÉED ARUGULA

Serves: 2
Prep Time: 10 minutes
Cook Time: 5 minutes

240 g arugula, washed and dried
15 ml extra virgin olive oil
Salt and pepper, to taste

Directions

1. In a large sauté pan, heat the olive oil over medium-high heat.
2. Add the arugula leaves to the pan and toss with the olive oil until just wilted, about 2-3 minutes.
3. Season with salt and pepper, to taste.
4. Serve immediately.

Notes

For added flavor, sprinkle with crushed red pepper flakes or grated lemon zest before serving.

AIR FRIED ARUGULA CHIPS

Serves: 1
Prep Time: 10 minutes
Cook Time: 5 minutes

120 g arugula, washed and dried
15 ml extra virgin olive oil
Salt and pepper, to taste

Directions

1. Preheat the air fryer to 190°C (375°F).
2. In a large mixing bowl, toss the arugula leaves with the olive oil, salt and pepper.
3. Spread the arugula leaves in a single layer on aluminum foil in the air fryer basket and place stainless steel utensils on top of the leaves to prevent movement.
4. Air fry for 3-5 minutes, or until the leaves are crispy and slightly browned.
5. Remove the arugula chips from the air fryer basket and transfer them to a serving platter.
6. Serve immediately as a snack or side dish or use as a topping for soups or salads.

Notes

Experiment with different seasonings, such as garlic powder or smoked paprika, for a unique twist on this healthy snack.

ASPARAGUS

History Asparagus, a perennial vegetable, has been valued for its succulent taste and medicinal properties since ancient times. The origins of asparagus can be traced back to the eastern Mediterranean region, where it was cultivated by the Greeks and Romans. As its popularity spread across Europe, it gained a reputation as a delicacy, particularly during the Renaissance period.

Can you eat it raw? Yes, you can eat asparagus raw. It has a fresh, earthy flavor and a crisp, juicy texture. However, cooking asparagus brings out its natural sweetness and tenderness, making it even more enjoyable.

Growing and harvesting Asparagus thrives in well drained, fertile soil with a pH of 6.0-8.0. It is typically grown from crowns, which are planted in early spring. The plant takes about 2-3 years to mature fully, after which it can be harvested annually for up to 15 years. Harvesting involves snapping off or cutting the spears at ground level when they reach a height of 6-8 inches.

Ripeness When selecting asparagus, choose spears that are firm, straight, and bright green. The tips should be tightly closed, and the stalks should snap easily when bent.

Spoilage Spoiled asparagus may develop a strong, unpleasant odor, sliminess, or mushiness. The spears will also lose their vibrant green color, turning yellow or brown.

Storage To store asparagus, wrap the ends of the stalks in a damp paper towel and place them in a plastic bag. Store the bag in the refrigerator's crisper drawer for up to a week.

Preserving The simplest way to preserve asparagus is by blanching and freezing. Blanch the spears in boiling water for 2-4 minutes, then transfer them to an ice bath to stop the cooking process. Drain, pat dry, and freeze in a single layer on a baking sheet before transferring to a freezer safe container.

PAN ROASTED ASPARAGUS

Serves: 2
Prep Time: 5 minutes
Cook Time: 12 minutes

250 g asparagus, tough ends trimmed
15 ml extra virgin olive oil
Fresh lemon juice (optional)
Salt and pepper, to taste

Directions

1. Heat a large skillet over a medium heat.
2. Once skillet is hot, add enough asparagus so there is a single layer.
3. Cover the skillet with a lid and cook until bottom side begins to char.
4. Drizzle the olive oil, salt, and pepper over the asparagus and toss to coat evenly.
5. Continue to cook the aspargus until al dente.
6. Remove from the pan and drizzle with lemon juice.
7. Serve hot or chilled

Notes

For added flavor, sprinkle some grated Parmesan cheese or crushed garlic over the asparagus before roasting.

ASPARAGUS SOUP

Serves: 4
Prep Time: 10 minutes
Cook Time: 20 minutes

450 g asparagus, tough ends trimmed and cut into pieces
30 ml extra virgin olive oil
1 L water or stock
Salt and pepper, to taste
Balsamic vinegar reduction (optional)

Directions

1. Heat the olive oil in a large pot over medium heat.
2. Add the asparagus and liquid to the pot and bring to a boil.
3. Reduce the heat and simmer until the asparagus is tender, about 10-15 minutes.
4. Using an immersion blender or a regular blender, blend the soup until smooth.
5. Season with salt and pepper.
6. Serve the soup hot with a drizzle of the balsamic reduction.

Notes

For a creamy texture, add a splash of heavy cream or coconut milk before blending the soup.

AIR FRIED ASPARAGUS

Serves: 2
Prep Time: 10 minutes
Cook Time: 7 minutes

250 g asparagus, tough ends trimmed
30 ml extra virgin olive oil
Salt and pepper, to taste

Directions

1. Preheat the air fryer to 200°C (400°F).
2. In a large mixing bowl, toss the asparagus with the olive oil until coated.
3. Season with salt and pepper.
4. Place the asparagus in the air fryer basket and cook for 5-7 minutes, shaking the basket occasionally, until the asparagus is crispy and tender.
5. Remove from the air fryer and serve hot as a side dish or snack.

Notes

For a crunchy coating, dip the asparagus in beaten egg and roll in breadcrumbs before air frying.

BEETROOT

History Beetroot has a fascinating history that dates back to ancient civilizations. The vegetable was initially cultivated primarily for its leaves, which were consumed for their medicinal properties. The Greeks and Romans also used beetroot as a natural dye for clothing and cosmetics. However, it was not until the 16th century that beetroot gained popularity as a food source. By the 19th century, beetroot had become a staple in various cuisines around the world, prized for its earthy taste, versatility, and vibrant color.

Can you eat it raw? Yes, beetroot can be enjoyed raw. When consumed in its raw form, beetroot has a crisp texture and a slightly sweet, earthy flavor. Raw beetroot is often thinly sliced or grated and used in salads, sandwiches, and as a garnish.

Growing and harvesting Beetroot is a cool season crop that grows best in well draining, fertile soil. The seeds are sown in spring or autumn and take around 50-70 days to reach maturity. Beetroot plants are usually harvested when the roots are 1.5 to 3 inches in diameter. However, the leaves can also be harvested throughout the growing season and used in a variety of dishes.

Ripeness To determine the ripeness of a beetroot, check the size of the root. It should be firm, smooth, and between 1.5 to 3 inches in diameter. Larger beetroots may be tougher and have a more fibrous texture. Additionally, the skin of the beetroot should be smooth and free from blemishes.

Spoilage Spoiled beetroots will become soft, wrinkled, and may develop mold or an off-putting smell. To prevent spoilage, store beetroots in a cool, dark place, and check them regularly for signs of deterioration.

Storage To store beetroots, cut off the leaves, rinse them, and put them in a bag with small holes or wrap them loosely in paper towels. Store them in the vegetable drawer of your refrigerator and check them often, throwing away any that are moldy or squishy.

Preserving The simplest way to preserve beetroots is by pickling. To pickle beetroots, boil them until tender, then peel and slice them. Prepare a pickling solution using water, vinegar, salt, and optionally, a bit of sugar. Place the beetroot slices in a sterilized jar and pour the pickling solution over them. Seal the jar and store in a cool, dark place.

BEETROOT SALAD

Serves: 2
Prep Time: 15 minutes
Cook Time: 0 minutes

1 medium beetroots, washed and peeled
15 ml extra virgin olive oil
15 ml fresh lemon juice
Salt and pepper, to taste

Directions

1. Grate the beetroots using a box grater or a food processor with a grating attachment, and transfer them to a large bowl.
2. Add the olive oil and lemon juice to the grated beetroots.
3. Season the mixture with salt and pepper, to taste.
4. Toss everything together until well combined.
5. Serve immediately or refrigerate until ready to serve.

Notes
To add more texture and flavor, you can add some chopped nuts or fresh herbs.

ROASTED BEETROOT

Serves: 1
Prep Time: 15 minutes
Cook Time: 40 minutes

1 medium beetroots, washed and peeled
15 ml extra virgin olive oil
Salt and pepper, to taste

Directions

1. Preheat your oven to 200°C (400°F).
2. Cut the beetroots into wedges and place them in a large bowl.
3. Add the olive oil, salt, and pepper to the bowl and toss well to ensure even coating.
4. Spread the seasoned beetroot wedges on a baking sheet in a single layer.
5. Roast in the preheated oven for 30-40 minutes, or until tender and slightly caramelized.
6. Serve hot.

Notes
Try roasting the whole peeled beet on a bed of course salt. You can also add some crumbled feta cheese or balsamic vinegar to the roasted beetroot for added flavor.

AIR FRIED BEET CHIPS

Serves: 1
Prep Time: 10 minutes
Cook Time: 15 minutes

1 medium beetroots, washed and peeled
30 ml extra virgin olive oil
Salt and pepper, to taste

Directions

1. Thinly slice the beetroots into even, thin rounds using a mandoline
2. In a large bowl, toss the beetroot slices with olive oil, salt, and pepper.
3. Arrange the seasoned beetroot slices in a single layer on aluminum foil in the air fryer basket, ensuring they do not overlap.
4. Air fry at 200°C (400°F) for 12-15 minutes, or until crispy, checking and shaking the basket occasionally to ensure even cooking.
5. Allow the beetroot chips to cool slightly before serving.

Notes
You can also experiment with different seasonings such as garlic powder, paprika, or dried herbs to add more flavor.

BELL PEPPER

History Bell peppers have a long and storied history dating back thousands of years. They are believed to have originated in South and Central America, where they were first domesticated by indigenous peoples. The ancient Mayans and Aztecs were known to cultivate and consume a variety of peppers for culinary and medicinal purposes. the Portuguese and Spanish explorers encountered peppers in the Caribbean and South America.

Can you eat it raw? Bell peppers can be enjoyed both raw and cooked. Raw peppers have a sweet, crisp, and refreshing taste, making them an excellent choice for salads, sandwiches, or as a standalone snack.

Growing and harvesting Bell peppers are warm season plants that thrive in full sunlight and well drained soil. They are typically planted in the spring and require a longer growing season compared to other pepper varieties, as they need to fully ripen on the plant. To promote healthy growth, it is essential to keep the soil consistently moist and fertilize the plants regularly. Harvesting occurs when the peppers have reached their full size and developed a deep, rich color.

Ripeness A ripe bell pepper has a vibrant red color, a firm texture, and a glossy appearance. Its skin should be taut and free of wrinkles. The pepper should feel heavy for its size, indicating that it is fully developed and ready to eat.

Spoilage A spoiled bell pepper may exhibit wrinkling or shriveling skin, soft spots, or mold growth. Additionally, the inside of the pepper may become slimy or develop an off smell. If any of these signs are present, it is best to discard the pepper.

Storing To maximize the shelf life of bell peppers, store them in the crisper drawer of your refrigerator, where they can last for up to two weeks. Ensure the peppers are unwashed and placed in a perforated plastic bag to maintain freshness.

Preserving The simplest way to preserve bell peppers is by freezing. First, wash and dry the peppers, then remove the seeds and membranes. Cut the peppers into your desired size and shape, and place them in a single layer on a baking sheet. Freeze the peppers for about an hour or until firm, then transfer them to a freezer safe container. Frozen peppers can be stored for up to 12 months.

SAUTÉED BELL PEPPER

Serves: 1
Prep Time: 5 minutes
Cook Time: 7 minutes

1 bell pepper
5 ml extra virgin olive oil
Salt and pepper, to taste

Directions

1. Wash the bell pepper, dry it, remove the seeds and membranes, and slice it into thin slices.
2. Heat olive oil in a sauté pan over medium heat.
3. When the oil is hot, add the red pepper slices, salt, and pepper to the pan and cook for 5-7 minutes, stirring occasionally, until the peppers are tender and caramelized.
4. Serve the sautéed peppers.

Notes
For a spicy kick, add a pinch of red pepper flakes or smoked paprika. Consider adding thinly sliced onions or garlic for extra flavor.

FIRE ROASTED BELL PEPPER

Serves: 1
Prep Time: 5 minutes
Cook Time: 30 minutes

1 bell pepper

Directions

1. Wash the bell pepper and dry it.
2. Secure the pepper over an open flame.
3. Roast each side of the pepper until fully charred.
4. Remove from flame and store in an airtight container for 5 minutes or until soft.
5. Remove the skin, seeds, and membrane from the pepper.
6. Use as desired.

Notes
Use tongs to handle the pepper once it is over the flame. Substitue for any fresh bell pepper recipes.

AIR FRIED BELL PEPPER

Serves: 1
Prep Time: 5 minutes
Cook Time: 10 minutes

1 bell pepper
15 ml extra virgin olive oil
Salt and pepper, to taste

Directions

1. Wash the bell pepper, dry it, remove the seeds and membranes, and slice it into thin slices.
2. Toss the red pepper slices with olive oil, salt, and pepper in a mixing bowl.
3. Preheat the air fryer to 200°C (400°F).
4. Place the red pepper slices in the air fryer basket and cook for 8-10 minutes, shaking halfway through, until the peppers are tender and slightly crispy.
5. Remove the air fried peppers from the air fryer and serve as a side dish or incorporate them into your favorite recipe.

Notes
For added flavor, sprinkle the red peppers with garlic powder or onion powder before air frying.

BOK CHOY

History Bok choy is a leafy green vegetable that belongs to the cruciferous vegetable family, which includes other nutrient dense vegetables such as kale, broccoli, and cabbage. It has been cultivated in Asia for over 5,000 years and is commonly used in Asian cuisine and traditional medicine for its numerous health benefits.

Can you eat it raw? Bok choy can be eaten raw, and its crisp texture and mild, slightly sweet flavor make it an excellent addition to salads or as a raw vegetable side dish. Raw bok choy leaves and stalks can also be used as a low carb wrap for sandwiches and other fillings.

Growing and harvesting Bok choy prefers cool weather and is typically grown during the spring and fall seasons in fertile, well draining soil. It requires consistent moisture to thrive, and gardeners should avoid overhead watering to prevent the spread of disease. Bok choy can be harvested when the leaves are tender and crisp, approximately 45 to 60 days after planting. Gardeners can either harvest the entire plant or just pick individual leaves as needed.

Ripeness To determine the ripeness of bok choy, look for firm, crisp stalks with vibrant green leaves.

Spoilage Spoiled bok choy exhibits limp, wilted leaves with yellow or brown discoloration, and the stalks may become slimy or emit an unpleasant odor. It is best to discard any bok choy showing these signs of spoilage.

Storing To store bok choy, wrap it loosely in a damp paper towel and place it inside a plastic bag, leaving the bag slightly open for air circulation. Store it in the crisper drawer of your refrigerator, where it can last for up to a week. Avoid washing bok choy before storage, as excess moisture can lead to spoilage.

Preserving The simplest way to preserve bok choy is by blanching and freezing. After washing and chopping the bok choy, briefly submerge it in boiling water for 30 seconds, then immediately transfer it to an ice bath to stop the cooking process. Pat dry and store it in airtight containers before placing it in the freezer.

SAUTÉED BOK CHOY

Serves: 1
Prep Time: 5 minutes
Cook Time: 5 minutes

1 bok choy, chopped
15 ml vegetable oil
Salt and pepper, to taste
Fresh lemon juice or rice vinegar (optional)

Directions

1. Heat the olive oil in a large pan over medium-high heat.
2. Add the chopped bok choy to the pan and sauté for 2-3 minutes, until slightly wilted.
3. Season with salt and pepper. Add the lemon juice or rice vinegar to the pan and stir to combine.
4. Serve immediately.

Notes
For a more flavorful dish, add minced garlic.

STEAMED BOK CHOY

Serves: 1
Prep Time: 5 minutes
Cook Time: 5 minutes

1 bok choy, chopped
5 ml sesame oil
Salt and pepper, to taste
Lime juice or white wine vinegar (optional)

Directions

1. Bring a pot of water to a boil and place a steamer basket on top.
2. Add the chopped bok choy to the steamer basket, cover, and steam for 3-5 minutes, until tender.
3. Remove the steamed bok choy from the basket and place it in a serving dish.
4. Drizzle the sesame oil over the bok choy.
5. Season with salt and pepper.
6. Add the lime juice or white wine vinegar to the dish and stir to combine.
7. Serve immediately.

Notes
For added flavor, top with toasted sesame seeds or crushed peanuts before serving.

AIR FRIED BOK CHOY

Serves: 1
Prep Time: 5 minutes
Cook Time: 6 minutes

1 bok choy, chopped
30 ml vegetable oil
Salt and pepper, to taste
Balsamic glaze (optional)

Directions

1. Preheat the air fryer to 200°C (400°F).
2. In a mixing bowl, toss the chopped bok choy with the oil and season with salt and pepper.
3. Place the seasoned bok choy in a single layer on aluminum foil in the air fryer basket.
4. Air fry the bok choy for 4-6 minutes, shaking the basket halfway through, until tender and slightly crispy.
5. Remove the air fried bok choy from the basket and place it in a serving dish.
6. Drizzle the balsamic over the bok choy.
7. Serve immediately.

Notes
For added texture, sprinkle the finished dish with toasted almonds or bacon.

BROCCOLI

History Broccoli is a cruciferous vegetable that is believed to have originated in Italy, specifically in the area of the eastern Mediterranean. It has been cultivated for over 2,000 years, with records showing its cultivation by the ancient Romans. Broccoli made its way to England and then to America in the 1700s.

Can you eat it raw? Yes, broccoli can be eaten raw. It is often consumed in salads, sandwiches, or as part of a vegetable tray. Raw broccoli has a crunchy texture and a slightly bitter, nutty flavor.

Growing and harvesting Broccoli is a cool season crop that grows best in temperatures between 60-70°F. It can be planted in the early spring or late summer for a fall harvest. Broccoli requires fertile, well draining soil and consistent moisture to thrive. The plants are typically harvested when the central head is firm, tight, and reaches 10-18 cm in diameter. It is important to harvest the central head before the flowers open and turn yellow.

Ripeness To determine the ripeness of broccoli, look for a firm, tightly packed head with a deep green or purplish green color. The florets should be tightly closed. The stem should be thick and firm.

Spoilage Spoiled broccoli will have a strong, unpleasant odor and may develop a slimy texture. The florets may become soft, mushy, and show signs of discoloration, such as yellowing or browning. If the broccoli has any of these signs, it should be discarded.

Storing Store unwashed broccoli in a loose, perforated plastic bag or wrapped in a damp paper towel in the refrigerator's crisper drawer. It will keep for up to a week. Do not wash the broccoli until you are ready to use it, as moisture can cause it to spoil more quickly.

Preserving The simplest way to preserve broccoli is by blanching and then freezing it. First, cut the broccoli into florets and blanch in boiling water for 2-3 minutes. Then, immediately transfer the florets to an ice bath to stop the cooking process. Once cooled, drain, pat dry, and freeze the florets in a single layer on a baking sheet before transferring them to a freezer safe container. Frozen broccoli can last up to 8 months in the freezer.

STEAMED BROCCOLI

Serves: 4
Prep Time: 10 minutes
Cook Time: 7 minutes

1 head of broccoli, cut into florets
15 ml of extra virgin olive oil
Fresh lemon juice (optional)
Salt and pepper, to taste

Directions

1. Fill a pot with 2.5 cm of water and bring it to a boil over high heat.
2. Place a steamer basket in the pot and add the broccoli florets.
3. Cover and steam for 5-7 minutes, or until the broccoli is tender but still crisp.
4. In a small bowl, whisk together the olive oil, lemon juice, salt, and pepper.
5. Transfer the steamed broccoli to a serving dish and drizzle the olive oil and lemon mixture over it.

Notes

To enhance the flavor, add grated Parmesan cheese or toasted pine nuts on top before serving.

SAUTÉED BROCCOLI

Serves: 4
Prep Time: 10 minutes
Cook Time: 7 minutes

1 head of broccoli, cut into florets
30 ml of extra virgin olive oil
Salt and pepper, to taste

Directions

1. Heat the extra virgin olive oil in a large skillet over medium heat.
2. Add the broccoli florets and sauté for 5-7 minutes, stirring occasionally, until tender but still crisp.
3. Season with salt and pepper to taste.

Notes

For added flavor, consider adding thinly sliced almonds or a sprinkle of grated Parmesan cheese.

AIR FRIED BROCCOLI

Serves: 4
Prep Time: 5 minutes
Cook Time: 12 minutes

1 head of broccoli, cut into florets
30 ml of extra virgin olive oil
Salt and pepper, to taste
Balsamic glaze (optional)

Directions

1. Preheat the air fryer to 200°C (400°F).
2. In a large bowl, toss the broccoli florets with extra virgin olive oil, salt, and pepper.
3. Place the florets in the air fryer basket and cook for 10-12 minutes, shaking halfway through, until tender and slightly browned.
4. Drizzle the reduced balsamic glaze over the air fried broccoli and serve.

Notes

For a more complex flavor, try adding a pinch of your favorite herbs or spices when seasoning the broccoli.

BRUSSELS SPROUTS

History Brussels sprouts are believed to have originated in ancient Rome and have been cultivated for over 400 years. The first written record of Brussels sprouts dates back to 1587 when they were mentioned in a book called "Hortus Romanus" by Joachim Camerarius. However, it was not until the 19th century that Brussels sprouts became more widely cultivated and popularized in Europe and the United States.

Can you eat it raw? Yes, Brussels sprouts can be eaten raw, but they are most commonly cooked by roasting, boiling, or steaming. Raw Brussels sprouts are crispy and have a slightly bitter taste, and can be eaten as a salad or added to coleslaw. It is important to note that some people may find the taste and texture of raw Brussels sprouts to be unappealing.

Growing and harvesting Brussels sprouts grow best in cool weather and are typically grown in the fall and winter months. They are typically planted in early spring and take around 80-100 days to reach maturity. The sprouts grow on a tall, thick stalk and can be harvested once they reach about 1-2 inches in diameter. When harvesting, it is best to pick the sprouts from the bottom of the stalk and work your way up.

Ripeness When choosing Brussels sprouts, look for firm, bright green sprouts with tight, compact leaves. Avoid sprouts with yellow or wilted leaves, or those that feel soft to the touch. The size of the sprouts doesn't necessarily indicate ripeness, as they will continue to grow as long as they remain on the stalk.

Spoilage Brussels sprouts can spoil quickly, so it is important to use them up within a few days of purchase. Signs that they have gone bad include a strong, unpleasant odor, mushy or discolored sprouts, or black spots on the leaves. It is best to discard any sprouts that show signs of spoilage to avoid the risk of foodborne illness.

Storing To store Brussels sprouts, keep them in a plastic bag in the refrigerator crisper drawer. They can be stored for up to 5 days. Avoid washing them until you are ready to use them, as excess moisture can cause them to spoil more quickly.

Preserving The simplest way to preserve Brussels sprouts is to blanch and freeze them. To do this, first, wash and trim the sprouts, then boil them in a large pot of salted water for 3-5 minutes. Remove the sprouts from the water and immediately plunge them into a bowl of ice water to stop the cooking process. Once cooled, drain the sprouts and pat them dry, then transfer them to a freezer safe container and freeze for up to 8 months.

PAN ROASTED BRUSSELS SPROUTS

Serves: 2
Prep Time: 5 minutes
Cook Time: 12 minutes

250 g Brussels sprouts, trimmed and halved
30 ml extra virgin olive oil
Balsamic glaze (optional)
Salt and pepper, to taste

Directions

1. Heat the olive oil in a large skillet over medium-high heat.
2. Add the Brussels sprouts to the skillet and cook, stirring occasionally, until they are browned and tender, about 10-12 minutes.
3. Add the balsamic and stir to coat the sprouts.
4. Season with salt and pepper, to taste.
5. Transfer the sprouts to a serving dish and serve hot.

Notes
For added flavor, consider incorperating garlic or shallots during the cooking process.

STEAMED BRUSSELS SPROUTS

Serves: 2
Prep Time: 5 minutes
Cook Time: 8 minutes

225 g Brussels sprouts, trimmed
45 ml water
Fresh lemon juice (optional)
Salt and pepper, to taste

Directions

1. Place the Brussels sprouts in a steamer basket set over a pot of boiling water.
2. Cover and steam until the sprouts are tender, about 6-8 minutes.
3. Remove the steamer basket from the pot and transfer the sprouts to a serving dish.
4. Drizzle the lemon juice over the sprouts and toss to coat.
5. Season with salt and pepper, to taste.
6. Serve hot.

Notes
For additional flavor, sprinkle some grated Parmesan cheese over the steamed Brussels sprouts before serving.

AIR FRIED BRUSSELS SPROUTS

Serves: 2
Prep Time: 5 minutes
Cook Time: 12 minutes

250 g Brussels sprouts, trimmed and qurtered
60 ml extra virgin olive oil
Salt and pepper, to taste

Directions

1. Preheat the air fryer to 400°F (200°C).
2. Toss the Brussels sprouts with olive oil and season with salt and pepper.
3. Place the sprouts on aluminum foil in the air fryer basket and cook for 10-12 minutes, shaking the basket occasionally, until the sprouts are crispy and slighlty charred.
4. Season with salt and pepper, to taste.
5. Transfer the sprouts to a serving dish and serve hot.

Notes
For a flavorful twist, toss the sprouts with roasted garlic oil or your favorite seasoning blend before air frying.

CABBAGE

History Cabbage has been cultivated for over 4,000 years and is believed to have originated in the eastern Mediterranean and Asia Minor regions. Ancient Greeks and Romans used cabbage for medicinal purposes and as a staple in their diets. In fact, the Greeks are credited with coining the name "krambe" for cabbage, which later evolved into the word "cabbage" we use today. During the Middle Ages, cabbage became a popular food crop throughout Europe and was widely used as a food source for sailors on long voyages due to its ability to withstand storage in harsh conditions.

Can you eat it raw? Cabbage can be eaten both raw and cooked, making it a versatile ingredient in many dishes. Raw cabbage is often used in salads, slaws, and sandwiches, providing a crisp and refreshing texture. The inner leaves of cabbage can also be used as a substitute for lettuce in tacos, wraps, and burgers.

Growing and harvesting Cabbage is typically grown in the cooler months, making it an ideal crop for fall and winter gardens. It can withstand light frosts, making it a popular choice for northern climates. Cabbage plants prefer fertile, well drained soil with a pH range of 6.0-6.5. They require consistent watering, especially during hot, dry weather. When harvesting cabbage, look for heads that are firm and heavy for their size, with crisp leaves that are tightly packed together.

Ripeness Cabbage is typically harvested when the heads are fully formed, which can take 70-100 days after planting depending on the variety. Signs of ripeness include firm, compact heads with a solid texture.

Spoilage Cabbage can spoil if not stored properly. Signs of spoilage include wilting, mold growth, and an unpleasant odor.

Storing To store cabbage, remove any damaged or wilted outer leaves, wrap the head in plastic wrap, and refrigerate. Cabbage can last up to two weeks in the refrigerator. If the cabbage is already cut, store it in an airtight container in the refrigerator for up to four days.

Preserving One of the simplest ways to preserve cabbage is to make sauerkraut. To do so, shred the cabbage, mix it with salt, and pack it tightly into a jar. Allow it to ferment at room temperature for several days to several weeks until it reaches your desired level of sourness. Sauerkraut can be stored in the refrigerator for several months.

ROASTED CABBAGE WEDGES

Serves: 6
Prep Time: 10 minutes
Cook Time: 30 minutes

1 head of cabbage, cut into wedges
30 ml extra virgin olive oil
15 ml apple cider vinegar (optional)
Salt and pepper, to taste

Directions

1. Preheat oven to 200°C (400°F).
2. In a bowl, whisk together the olive oil, vinegar, salt, and pepper.
3. Brush the cabbage wedges with the mixture, ensuring to coat all sides.
4. Place the wedges on a baking sheet and roast for 25-30 minutes, until tender and lightly browned.

Notes
Consider adding garlic powder, onion powder, or paprika to the mixture for additional flavor.

CABBAGE SLAW

Serves: 4
Prep Time: 15 minutes
Cook Time: 0

1/2 head of cabbage, thinly sliced
2 large carrots, shredded
30 ml extra virgin olive oil
30 ml fresh lemon or lime juice
Salt and pepper, to taste

Directions

1. In a large bowl, combine the sliced cabbage and shredded carrots.
2. In a separate bowl, emulsify together the olive oil, juice, salt, and pepper.
3. Pour the dressing over the cabbage and carrots and toss until evenly coated.
4. Refrigerate for at least 30 minutes before serving.

Notes
Add raisins, sunflower seeds, or chopped nuts for additional texture and flavor. Substitute oil and lemon juice for mayonaise for a deli style coleslaw.

AIR FRIED CABBAGE CHIPS

Serves: 4
Prep Time: 10 minutes
Cook Time: 15 minutes

1/2 head of cabbage, thinly sliced
60 ml extra virgin olive oil
Salt and pepper, to taste

Directions

1. Preheat the air fryer to 200°C (400°F).
2. In a bowl, toss the thinly sliced cabbage with olive oil, salt, and pepper.
3. Place the cabbage in the air fryer basket, ensuring it is in a single layer on aluminum foil.
4. Air fry for 10-15 minutes, shaking the basket halfway through, until the cabbage is crispy and lightly browned.

Notes
Experiment with different seasonings, such as garlic powder, onion powder, or chili powder, to customize the flavor.

CARROTS

History Carrots have a long and varied history, and their origins can be traced back to Central Asia, where wild carrots are still found today. The earliest known cultivation of carrots dates back over 5,000 years to the region that is now Afghanistan. These early carrots were not orange but rather purple, white, and yellow. It was not until the 16th century that orange carrots became more popular in Europe, where they were selectively bred for their color and flavor.

Can you eat it raw? Yes, carrots can be eaten raw and are a versatile vegetable that can be consumed in many ways. They are commonly used as a crunchy addition to salads or as a snack with dip. Raw carrots are also a great source of nutrition and can be enjoyed as part of a healthy diet.

Growing and harvesting Carrots are typically grown in loose, well-draining soil. They require consistent moisture to grow properly, and the soil should be kept moist but not waterlogged. Carrots are planted from seeds and can take several months to mature, depending on the variety. They are ready to be harvested when they have reached their full size and color, which varies depending on the variety. Carrots are typically harvested by pulling them out of the ground by their tops.

Ripeness Carrots are ready to be harvested when the tops of the roots are about 1 inch in diameter and the roots are a bright orange color. The roots should be firm and crisp to the touch. The flavor of the carrot will also be at its peak when fully ripe.

Spoilage When carrots have spoiled, they will become soft and mushy, and may have a bad odor. Mold may also be present on the surface of the carrot. It's important to inspect carrots carefully before consuming them, especially if they have been stored for a while.

Storing To store carrots, remove the tops and store them in a plastic bag in the refrigerator. Carrots can last for up to two weeks when stored properly. It's important to keep carrots away from ethylene-producing fruits and vegetables, such as apples and avocados, as this can cause them to spoil more quickly.

Preserving Carrots can be preserved by blanching them in boiling water for 2-3 minutes and then freezing them. This method helps to maintain the flavor and texture of the carrots, making them a great option for use in soups, stews, and other dishes.

ROASTED CARROTS

Serves: 2
Prep Time: 10 minutes
Cook Time: 25 minutes

250 g carrots, peeled and sliced pieces
15 ml extra virgin olive oil
Fresh lemon juice (optional)
Salt and pepper, to taste

Directions

1. Preheat the oven to 200°C (400°F).
2. In a bowl, toss the sliced carrots with olive oil, lemon juice, salt, and pepper until evenly coated.
3. Arrange the carrots in a single layer on a baking sheet.
4. Roast the carrots for 20-25 minutes or until tender and lightly browned.
5. Transfer the roasted carrots to a serving dish and serve immediately.

Notes

For added flavor, sprinkle some fresh chopped parsley or grated Parmesan cheese over the roasted carrots before serving.

CARROT SOUP

Serves: 4
Prep Time: 10 minutes
Cook Time: 30 minutes

450 g carrots, peeled and chopped
30 ml extra virgin olive oil
15 ml apple cider vinegar
500 ml water or stock
Salt and pepper, to taste

Directions

1. In a large pot, heat olive oil over medium heat.
2. Add the chopped carrots and sauté for 5-7 minutes or until lightly browned.
3. Add liquid, vinegar, salt, and pepper.
4. Bring the mixture to a boil, then reduce the heat and simmer for 20-25 minutes or until the carrots are very tender.
5. Remove the pot from the heat, cool
6. Using an immersion blender, puree the soup until smooth.

Notes

Add a dollop of sour cream or Greek yogurt to the soup for a tangy creaminess.

AIR FRIED CARROT FRIES

Serves: 2
Prep Time: 10 minutes
Cook Time: 12 minutes

250 g carrots, peeled and cut into fry sized pieces
45 ml extra virgin olive oil
Salt and pepper, to taste

Directions

1. Preheat the air fryer to 200°C (400°F).
2. In a bowl, toss the carrot fries with olive oil, salt, and pepper until evenly coated.
3. Place the carrot fries in the air fryer basket in a single layer on aluminum foil.
4. Air fry for 10-12 minutes, shaking the basket halfway through, until the carrot fries are tender and lightly browned.
5. Remove from the air fryer and serve.

Notes

To enhance flavor, consider adding paprika, garlic powder, or onion powder to the carrot fries before air frying.

CAULIFLOWER

History Cauliflower has a long and interesting history, believed to have originated in the Mediterranean and Asia Minor regions. Its cultivation can be traced back to the 6th century B.C., where it was grown in ancient Roman gardens. In the 16th century, cauliflower was introduced to Western Europe by the Italian scientist, Pietro Andrea Mattioli. It then spread to other parts of the world, including India, China, and the Americas.

Can you eat it raw? Yes, cauliflower can be eaten raw. It is often served raw as a crudité or as part of a salad. Raw cauliflower has a crisp texture and a slightly nutty, sweet taste. It can also be roasted, grilled, or used in stir-fries and soups.

Growing and harvesting Cauliflower grows best in cool temperatures between 60-65°F (15.5-18.3°C). It prefers fertile, well-drained soil and needs consistent watering. Cauliflower is typically planted in late spring to early summer, and the heads are ready to harvest 2-3 months later. To harvest cauliflower, the entire head is cut off the plant with a sharp knife, making sure to leave some of the leaves attached to protect the head.

Ripeness Cauliflower is ripe when the head is firm and compact, and the leaves surrounding the head are bright green and healthy. The head should be a creamy white color, with no signs of yellowing or browning. If the cauliflower head has begun to spread open, it is likely overripe.

Spoilage Spoiled cauliflower will have a sour, rancid smell and may have dark spots or discoloration on the head. The head may also be soft to the touch and have a slimy texture. When in doubt, it is best to discard cauliflower that appears spoiled.

Storing Store cauliflower in the refrigerator, wrapped in a plastic bag, for up to a week. If the cauliflower has been cut or cooked, store it in an airtight container in the refrigerator and consume it within 2-3 days. Do not wash cauliflower before storing, as excess moisture can cause it to spoil faster.

Preserving Cauliflower can be frozen for long-term storage. To freeze cauliflower, wash and cut it into florets, blanch for 3 minutes, then transfer to a colander and rinse with cold water. Pat the cauliflower dry with a paper towel, then spread it out on a baking sheet and place in the freezer until frozen. Once frozen, transfer the cauliflower to a freezer safe container store in the freezer for up to 8 months.

ROASTED CAULIFLOWER

Serves: 4
Prep Time: 10 minutes
Cook Time: 25 minutes

1 head of cauliflower, cut into florets
45 ml exrta virgin olive oil
Balsamic reduction (optional)
Salt and pepper to taste

Directions

1. Preheat the oven to 200°C (400°F).
2. In a large bowl, toss the cauliflower florets with the olive oil, balsamic, salt, and pepper.
3. Spread the cauliflower evenly on a lined baking sheet in a single layer.
4. Roast the cauliflower for 20-25 minutes, or until golden brown and tender.
5. Serve hot.

Notes

For added flavor, consider adding minced garlic and a sprinkle of your favorite herbs, such as parsley or thyme, to the cauliflower before roasting.

CARAMELIZED CAULIFLOWER

Serves: 4
Prep Time: 10 minutes
Cook Time: 1-3 hours

1 head of cauliflower, chopped
45 ml extra virgin olive oil
Salt and pepper, to taste

Directions

1. Heat the olive oil in a large pot over medium heat.
2. Add the cauliflower and cook on medium until the florets start to caramalize.
3. Reduce the heat and cook until the cauliflower florets completly lose structure and stir as often as needed to prevent burning.
4. Use spoon to mash or an immersion blender to puree the cauliflower.
5. Season with the salt and pepper, to taste.
6. Serve hot.

Notes

Add some fresh herbs such as thyme or rosemary. For a creamier texture, stir in a splash of heavy cream or coconut milk before serving

AIR FRIED CAULIFLOWER

Serves: 2
Prep Time: 10 minutes
Cook Time: 15 minutes

1/2 head of cauliflower, cut into florets
60 ml extra virrgin olive oil
Salt and pepper, to taste

Directions

1. Preheat the air fryer to 200°C (400°F).
2. In a bowl, toss the cauliflower florets with the olive oil, salt, and pepper.
3. Place the cauliflower in the air fryer basket on aluminum foil and cook for 12-15 minutes, shaking the basket every 5 minutes, or until golden brown and tender.
4. Remove the cauliflower from the air fryer and serve hot.

Notes

For extra flavor, add a pinch of red pepper flakes or smoked paprika to the cauliflower before air frying.

CELERY

History Celery has a rich history that dates back to ancient times. It is a member of the Apiaceae family and is believed to have originated in the Mediterranean region. The ancient Greeks and Romans used celery as a medicinal herb and for its aromatic qualities in cooking. In fact, celery was so highly valued that it was used as a prize for winners of athletic games in ancient Greece. It was not until the 17th century that celery began to be cultivated and appreciated as a food item.

Can you eat it raw? Yes, celery is commonly eaten raw and is a popular snack food. It can also be used as a crunchy garnish for soups and salads.

Growing and harvesting Celery is a cool season crop that requires fertile soil and consistent moisture. It is typically grown as an annual plant, and is usually started indoors as seedlings before being transplanted to the garden once they have several leaves. Celery plants need to be blanched to prevent them from turning bitter. This is done by covering the stalks with cardboard or paper collars, which shield them from sunlight. Celery is harvested by cutting the entire plant at ground level.

Ripeness When choosing celery, look for firm, tight stalks that are a pale green color. The leaves should be fresh looking and not wilted. Avoid celery that has soft or discolored spots.

Spoilage Spoiled celery will have a strong odor and may be slimy or have mold growing on it. If you notice any signs of spoilage, discard the celery.

Storing To keep celery fresh, store it in the refrigerator in a plastic bag or wrap it in aluminum foil. It will stay fresh for up to two weeks. If the celery starts to wilt, you can revive it by placing it in a bowl of cold water for a few hours.

Preserving The simplest way to preserve celery is to freeze it for later use. Cut the celery into small pieces, blanch them in boiling water for 3 minutes, and then freeze them in a single layer on a baking sheet. Once frozen, transfer the celery to a freezer safe container.

SAUTÉED CELERY

Serves: 1
Prep Time: 5 minutes
Cook Time: 7 minutes

125 g sliced celery
5 ml extra virgin olive oil
Salt and pepper, to taste

Directions

1. Heat the oil in a skillet over medium heat.
2. Add the sliced celery and sauté for 5-7 minutes, until tender.
3. Season with salt and pepper, to taste.
4. Serve hot.

Notes

For a twist on this recipe, you can add sliced onions or garlic to the skillet while sautéing the celery.

CELERY SALAD

Serves: 2
Prep Time: 10 minutes
Cook Time: 0

225 g celery, chopped
15 ml extra virgin olive oil
15 ml lemon juice
Salt and pepper, to taste

Directions

1. Place the celery pieces in a bowl.
2. Add the olive oil, lemon juice, salt, and pepper.
3. Toss to combine.
4. Serve chilled.

Notes

To remove celery fibers use a vegetable peeler. You can also add additional vegetables or fruits, such as cherry tomatoes or apples, to the salad for extra flavor and nutrition.

AIR FRIED CELERY FRIES

Serves: 2
Prep Time: 5 minutes
Cook Time: 12 minutes

225 g celery cut into fries
30 ml vegetable oil
Salt and pepper, to taste

Directions

1. Preheat the air fryer to 200°C (400°F).
2. Toss the celery pieces with the oil, salt, and pepper.
3. Place the celery in the air fryer basket on aluminum foil and cook for 5-7 minutes or until they are crispy and tender.
4. Serve hot with your favorite dipping sauce.

Notes

You can also experiment with different seasonings, such as paprika or garlic powder, for added flavor.

CHILI PEPPER

History Chili peppers have a long and fascinating history that dates back thousands of years. It is believed that chili peppers were first domesticated in Mexico over 6,000 years ago and were used extensively by the ancient Mayans and Aztecs in their cooking. The popularity of chili peppers eventually spread throughout the world, thanks in part to Spanish and Portuguese explorers who brought them to Asia, Europe, and Africa in the 15th and 16th centuries.

Can you eat it raw? Chili peppers can certainly be eaten raw, but their spiciness may be too much for some people to handle. The heat level of chili peppers can be reduced by removing the seeds and the white membrane inside, which contains most of the capsaicin that gives them their heat.

Growing and harvesting Chili peppers are relatively easy to grow and can be cultivated in warm and sunny climates around the world. They can be grown from seeds and require well draining soil, regular watering, and fertilization. Chili peppers are typically harvested when they are fully mature and have reached their desired color, which varies depending on the variety.

Ripeness The ripeness of chili peppers can be determined by their color. Most chili peppers start out green and then change color as they mature. For example, jalapeño peppers start green, then turn red when fully ripe. The longer they are left on the plant, the hotter they become.

Spoilage Chili peppers can spoil quickly, so it is important to keep an eye on them. Signs of spoilage include mold, soft spots, and a slimy texture. If you notice any of these signs, it is best to discard the chili pepper.

Storing Fresh chili peppers can be stored in the refrigerator for up to two weeks. It is best to store them in a plastic bag with a paper towel to absorb any excess moisture. Dried chili peppers can be stored in an airtight container in a cool, dark place for up to a year.

Preserving
Drying is one of the simplest ways to preserve chili peppers. To dry chili peppers, simply hang them in a dry and well ventilated area until they are fully dehydrated. Once dried, they can be stored in an airtight container for up to a year.

GRILLED CHILI PEPPERS

Serves: 1
Prep Time: 5 minutes
Cook Time: 6 minutes

4 chili peppers
15 ml vegetable oil
Fresh lime juice (optional)
Salt and pepper, to taste

Directions
1. Preheat the grill to medium-high heat.
2. Cut the chili peppers in half and remove the seeds and membrane.
3. In a small bowl, whisk together oil, lime juice, salt, and pepper.
4. Brush the oil mixture onto the chili peppers.
5. Place the chili peppers on the grill and cook for 2-3 minutes on each side or until they are charred and tender.

Notes
Experiment with different types of chili peppers for varying heat levels. Serve with a side of yogurt or sour cream to balance the spiciness.

CHILI PEPPER VINAIGRETTE

Serves: 2
Prep Time: 10 minutes
Cook Time: 0

2 chili peppers, sliced thinly
30 ml extra virgin olive oil
15 ml apple cider vinegar
Salt and pepper, to taste

Directions
1. In a large bowl, immersion blend together oil, apple cider vinegar, salt, and pepper.
2. Add the chili peppers to the bowl and toss until everything is coated with the dressing, blend untill smooth if desired
3. Serve immediately.

Notes
For added crunch and flavor, add toasted nuts or seeds to the salad. You can try using grilled chili peppers for a deeper flavor profile.

AIR FRIED CHILI PEPPERS

Serves: 2
Prep Time: 5 minutes
Cook Time: 7 minutes

4 chili peppers
15 ml vegetable oil
Salt and pepper, to taste

Directions
1. Preheat the air fryer to 190°C (375°F).
2. Cut the chili peppers in half and remove the seeds and membrane.
3. Brush the chili peppers with oil and sprinkle with salt and pepper.
4. Place the chili peppers in the air fryer basket on aluminum foil and cook for 5-7 minutes or until they are crispy and tender.

Notes
Serve the air fried chili peppers with a dipping sauce of your choice, such as aioli or ranch. You can also stuff them with cheese before air frying for a creamy, spicy treat.

COLLARD GREENS

History Collard greens have a rich history dating back thousands of years. The exact origin of collard greens is uncertain, but they are believed to have originated in the eastern Mediterranean region. They were brought to the United States by African slaves during the slave trade, and quickly became a staple in Southern cuisine.

Can you eat it raw? While collard greens are often cooked, they can also be eaten raw. Young, tender collard greens can be eaten raw in salads, sandwiches, or wraps. The older, tougher leaves are best cooked before eating.

Growing and harvesting Collard greens are a cool season crop that is typically grown in the early spring or late summer. They grow best in full sun to partial shade and in well drained soil that is rich in organic matter. Collard greens can be harvested when the leaves are large and dark green, usually around 60 to 75 days after planting. It is important to pick the leaves regularly to encourage continued growth and to prevent the plant from bolting or going to seed.

Ripeness When selecting collard greens, look for leaves that are dark green and free of yellow spots or holes. The leaves should be crisp and firm, not wilted or limp. Young collard greens tend to be more tender and have a milder flavor, while older leaves can be tougher and more bitter. However, older leaves are still edible and can be cooked down into stews or soups.

Spoilage Collard greens can spoil quickly if they are not stored properly. Look for signs of mold or discoloration, and discard any leaves that are slimy or foul smelling. Also, avoid buying collard greens that have been precut, as they are more susceptible to spoilage.

Storing To keep collard greens fresh, remove any bands or ties, wrap them in a damp paper towel, and place them in a plastic bag in the refrigerator. They can be stored for up to a week this way.

Preserving The simplest way to preserve collard greens is to freeze it for later use. Place the leaves in an airtight container and store them in the freezer for up to 8 months.

SAUTÉED COLLARD GREENS

Serves: 4
Prep Time: 10 minutes
Cook Time: 7 minutes

1 bunch collard greens
15 ml extra virgin olive oil
Salt and pepper, to taste
Fresh lemon juice or balsamic glaze (optional)

Directions

1. Rinse the collard greens and remove the tough stems. Cut the leaves into bite sized pieces.
2. Heat the oil in a large skillet over medium heat.
3. Add the collard greens, salt, and pepper to the skillet.
4. Sauté the collard greens for 5-7 minutes, or until the leaves are wilted and tender.
5. If desired, add a squeeze of lemon juice or drizzle of balsamic.
6. Serve warm as a side dish.

Notes
To enhance flavor, consider adding minced garlic or chopped onions during the sautéing process.

COLLARD GREEN SALAD

Serves: 4
Prep Time: 10 minutes
Cook Time: 0 minutes

1 bunch collard greens
30 ml extra virgin olive oil
15 ml fresh lemon juice
Salt and pepper, to taste

Directions

1. Rinse the collard greens and remove the tough stems. Cut the leaves into bite sized pieces.
2. In a small bowl, whisk together the oil, lemon juice or vinegar, salt, and pepper.
3. Toss the collard greens with the dressing in a large bowl.
4. Serve immediately as a fresh and healthy salad.

Notes
For added texture and flavor, consider adding chopped nuts, dried cranberries, or crumbled feta cheese to the salad.

AIR FRIED COLLARD GREENS CHIPS

Serves: 2
Prep Time: 10 minutes
Cook Time: 10 minutes

1 bunch collard greens
30 ml vegetable oil
Salt and pepper, to taste

Directions

1. Preheat the air fryer to 175°C (350°F).
2. Rinse the collard greens and pat them dry. Remove the tough stems and tear the leaves into small pieces.
3. In a bowl, toss the collard greens with the oil, salt, and pepper.
4. Arrange the collard greens in a single layer in the air fryer basket on aluminum foil.
5. Air fry the collard greens for 5-10 minutes, or until crispy.
6. Serve as a healthy snack or garnish for soups and salads.

Notes
Experiment with different seasonings to create a variety of flavors for your collard green chips.

CORN

History Corn, also known as maize, is one of the most widely cultivated and consumed crops in the world. It has been an important staple food for indigenous peoples in the Americas for thousands of years. Corn was first domesticated in what is now Mexico around 10,000 years ago, and it spread throughout the Americas, becoming an essential part of many cultures and cuisines. When European explorers arrived in the Americas, they brought corn back to Europe, where it eventually became an important crop in many countries.

Can you eat it raw? Yes, corn can be eaten raw, although it is more commonly cooked. Raw corn can be eaten off the cob or shaved off and added to salads, salsas, or other dishes for a crunchy texture.

Growing and harvesting Corn is typically grown in warm climates with plenty of sunlight and water. It requires well drained soil and regular fertilization. The plants are usually started from seeds, which are sown in rows or hills. Corn is ready to be harvested when the kernels are plump and tender, and the silks at the end of the ear have turned brown. The exact time of harvesting depends on the variety of corn and the climate in which it is grown.

Ripeness To tell if corn is ripe, look for plump, juicy kernels that are tightly packed together. The husk should be green and slightly moist, and the silks at the end of the ear should be dry and brown. If the kernels are not fully developed, the corn is not yet ripe.

Spoilage Spoiled corn will have discolored, mushy, or moldy kernels. It may also have an unpleasant odor. When buying corn, look for ears that are firm and free of blemishes or soft spots.

Storing Fresh corn should be stored in the refrigerator, preferably still in the husk, until ready to use. This will help to preserve its freshness and flavor. Cooked corn can be stored in an airtight container in the refrigerator for up to 4 days.

Preserving The simplest way to preserve corn is to freeze it for later use. To do this, blanch the ears of corn in boiling water for 3-4 minutes, then transfer them to a bowl of ice water to stop the cooking process. Once cooled, cut the kernels from the cob and store them in an airtight container in the freezer for up to 6 months.

ROASTED CORN

Serves: 1
Prep Time: 5 minutes
Cook Time: 45 minutes

1 ear of corn
15 ml melted butter or vegetable oil
Fresh lime juice (optional)
Salt and pepper, to taste

Directions

1. Preheat the oven to 200°C (400°F).
2. Place the ears of corn, still in their husks, directly on the oven rack.
3. Roast for 35-45 minutes, or until the kernels are tender and the husks are slightly charred.
4. Remove from the oven and let cool for a few minutes.
5. Remove the husks and silk from the corn.
6. Cut the kernels from the cob and place in a bowl.
7. Drizzle with butter or oil, lime juice, salt, and pepper. Toss to coat.
8. Serve warm.

Notes

For added flavor, you can sprinkle the corn with some chili powder or smoked paprika before serving.

GRILLED CORN

Serves: 1
Prep Time: 10 minutes,
Cook Time: 12 minutes

1 ear of corn
15 ml melted butter or extra virgin olive oil
Balsamic glaze or fresh lime juice (optional)
Salt and pepper, to taste

Directions

1. Preheat the grill to medium-high heat.
2. Brush the ears of corn with oil or but.
3. Grill the corn for 10-12 minutes, turning occasionally, until slightly charred and tender.
4. Remove from the grill and let cool for a few minutes.
5. (Optional) Cut the kernels from the cob and place in a bowl.
6. Cover the corn in the butter or olive oil, balsamic or lime, salt, and pepper.

Notes

Cover grilled corn with a mixture of sour cream, mayonnaise, lime juice, chili powder, and paprika then top with cotija cheese and cilantro.

AIR FRIED CORN ON THE COB

Serves: 1
Prep Time: 7 min
Cook Time: 15 minutes

1 ear of corn
30 ml melted butter or vegetable oil
Fresh lemon juice (optional)
Salt and pepper, to taste

Directions

1. Clean the ears of corn of the husk and fibers, wash, and then pat dry.
2. Cut ears of corn in half.
3. Coat the corn so every kernel is covered with the oil or butter, and set aside excess.
4. Arrange the corn halves in a single layer in the air fryer basket.
5. Air fry the corn at 200°C (400°F) for 10-15 minutes, or until golden.
6. Immediately sprinkle with salt and pepper and serve.

Notes

For an extra burst of flavor, drizzle the cooked corn with the optional lemon juice before serving.

CUCUMBER

History Cucumbers are native to South Asia and have been cultivated for over 3,000 years. They were highly valued in ancient civilizations such as Egypt, Greece, and Rome, where they were often used for skincare, believed to have healing properties, and consumed for their taste and cooling effect.

Can you eat it raw? Yes, cucumbers are often eaten raw and are commonly used in salads, sandwiches, and as a refreshing snack.

Growing and harvesting Cucumbers are warm season vegetables that thrive in well draining soil with adequate moisture and nutrients. They require regular watering and warm temperatures for successful growth. Cucumbers can be grown from seeds, and it is recommended to sow them directly into the garden after the threat of frost has passed. Trellising or providing some support helps in upward growth and prevents disease. Harvesting can begin when the fruit is about 6-8 inches long and the skin is still firm and bright green.

Ripeness Cucumbers are usually harvested when they are immature, and the skin is a bright green color. Mature cucumbers may turn yellow and develop a bitter taste.

Spoilage Cucumbers can spoil quickly if not stored properly. Signs of spoilage include soft or mushy areas, mold growth, and a sour smell.

Storage To keep cucumbers fresh, store them in the refrigerator in a plastic bag or wrapped in a damp cloth to prevent dehydration. It is recommended to keep them away from fruits and vegetables that produce ethylene gas, such as tomatoes and bananas, as this can cause them to spoil faster.

Preservation Pickling is a popular method of preserving cucumbers. Cucumbers specifically for pickling are smaller, have a thicker skin, and fewer seeds than regular cucumbers, making them ideal for pickling. They can be preserved using vinegar, salt, and various spices to create a flavorful and crunchy pickle.

CUCUMBER SALAD

Serves: 2
Prep Time: 10 minutes
Cook Time: 0

1 cucumbers, sliced
30 ml extra virgin olive oil
15 ml apple cider vinegar
Salt and pepper, to taste

Directions

1. In a bowl, combine the sliced cucumbers, olive oil, vinegar, salt, and pepper.
2. Mix until cucumbers are well coated.
3. Chill in the refrigerator for 15 minutes before serving.

Notes:
For added flavor, consider incorporating sour cream or fresh herbs such as dill or parsley. To make the salad more filling, add sliced red onion and crumbled feta cheese.

CUCUMBER SOUP

Serves: 4
Prep Time: 10 minutes
Cook Time: 0

2 cucumbers, peeled and chopped
20 ml extra virgin olive oil
15 ml lemon juice
Salt and pepper to taste

Directions

1. Using an imersion blender, puree the chopped cucumbers, olive oil, lemon juice, salt, and pepper until completely smooth.
2. Chill in the refrigerator for 30 minutes before serving.

Notes
For a creamier soup, add a dollop of Greek yogurt or sour cream. You can also garnish with chopped fresh herbs like mint, cilantro, or dill for added flavor.

AIR FRIED PICKLED CUCUMBER CHIPS

Serves: 3
Prep Time: 5 minute
Cook Time: 12 minutes

1 whole freshly pickled cucumbers, sliced
30 ml frying oil
Salt and pepper, to taste

Directions

1. Preheat the air fryer to 200°C (400°F).
2. In a bowl, toss the sliced cucumbers with olive oil, salt, and pepper until well coated.
3. Place the cucumber slices in a single layer on aluminum foil in the air fryer basket and cook for 10-12 minutes, flipping halfway through cooking.
4. Serve hot.

Notes
For added crunch, coat the cucumber slices in a beaten egg mixture and seasoned breadcrumbs before air frying. Experiment with different seasonings, such as garlic powder or smoked paprika, to customize the flavor.

EGGPLANT

History Eggplants have a long history of cultivation, with evidence suggesting that they were first domesticated in India over 4,000 years ago. From there, they spread to other parts of Asia, including China, where they were first recorded in the 5th century. The Arab world also played a significant role in the spread of eggplants, with the vegetable making its way to the Mediterranean region in the Middle Ages. It was during this time that the vegetable was given the name "aubergine," which is still commonly used in many parts of the world. Eggplants were introduced to Europe in the 16th century, and they became popular in the cuisine of many countries, including Italy, Greece, and Spain.

Can you eat it raw? Eggplant is generally not eaten raw due to its bitter taste and spongy texture. However, some people enjoy thinly sliced raw eggplant marinated in a vinegar based dressing or as a garnish on salads.

Growing and harvesting Eggplants are warm weather plants that require a long growing season. They are typically grown from seedlings and need well draining soil with a pH between 5.5 and 6.5. Eggplants are ready to harvest when they are firm and shiny, and their skin is smooth and unblemished. They should be harvested before the seeds inside mature, as this can cause the eggplant to become bitter.

Ripeness A ripe eggplant should be firm to the touch and have a smooth, unblemished skin. The flesh should be tender and creamy, with small, soft seeds. Avoid eggplants that are wrinkled or have brown spots, as these are signs of over-ripeness.

Spoilage Eggplants can spoil quickly, so it is important to use them as soon as possible after harvesting or purchasing. Signs of spoilage include a mushy texture, discolored skin, and a sour smell.

Storing Eggplants can be stored in the refrigerator for up to five days. They should be wrapped in paper towels and placed in a plastic bag to prevent moisture buildup.

Preserving The simplest way to preserve eggplants is to roast or grill them and then freeze them in an airtight container for up to six months. They can also be pickled or canned for long term storage.

GRILLED EGGPLANT

Serves: 4
Prep Time: 10 minutes
Cook Time: 10 minutes

1 eggplant, sliced into rounds
60 ml extra virgin olive oil
Balsamic glaze (optional)
Salt and pepper, to taste

Directions

1. Preheat a grill or grill pan to medium-high heat.
2. Brush the eggplant slices with olive oil and season with salt and pepper.
3. Grill the eggplant slices for 4-5 minutes on each side, until they are tender and slightly charred.
4. Remove the eggplant from the grill and drizzle with balsamic.
5. Serve immediately.

Notes

For an added touch of flavor, sprinkle some fresh chopped parsley or grated Parmesan cheese on top before serving.

ROASTED EGGPLANT SALAD

Serves: 4
Prep Time: 10 minutes
Cook Time: 30 minutes

1 eggplant, diced
60 ml extra virgin olive oil
15 ml balsamic glaze
Salt and pepper, to taste

Directions

1. Preheat the oven to 190°C (375°F).
2. Toss the diced eggplant with olive oil and season with salt and pepper.
3. Roast the eggplant in the oven for 25-30 minutes, until it is tender and slightly caramelized.
4. In a separate bowl, whisk together the red wine vinegar and remaining olive oil.
5. Toss the roasted eggplant with the dressing.
6. Serve the salad immediately or refrigerate until ready to serve.

Notes

Add some crumbled feta cheese or sliced olives for extra flavor and texture.

AIR FRIED EGGPLANT

Serves: 4
Prep Time: 10 minutes
Cook Time: 10 minutes

1 eggplant, sliced into thin rounds
40 ml extra virgin olive oil
Salt and pepper, to taste

Directions

1. Preheat an air fryer to 200°C (400°F).
2. In a shallow dish, mix together the breadcrumbs, salt, and pepper.
3. Dip the eggplant slices in the olive oil, then place the coated eggplant slices in the air fryer basket in a single layer on aluminum foil.
4. Air fry the eggplant slices for 8-10 minutes, until they are crispy and golden brown.
5. Serve the eggplant Parmesan immediately.

Notes

Add grated Parmesan cheese to breadcrumb mixture. Serve with marinara sauce for dipping and a side salad for a complete meal.

GARLIC

History Garlic is believed to have been first domesticated over 5000 years ago in central Asia, and its use as both a food ingredient and medicinal herb dates back to ancient times. It was highly prized by the Egyptians, who used it in food preparation, medicine, and religious rituals.

Can you eat it raw? Garlic can indeed be eaten raw, and it is often used in dishes like bruschetta, salsa, and salad dressings to add a pungent, sharp flavor. Raw garlic has a stronger and sharper taste than cooked garlic, and it can also have a more pungent aroma. However, the high concentration of compounds like allicin in raw garlic can also cause digestive discomfort in some people.

Growing and harvesting Garlic is typically grown from cloves, which are the individual segments of the garlic bulb. The cloves are planted in the fall, and the garlic bulb grows underground during the winter months. In the summer, when the leaves start to turn yellow and wilt, the garlic bulbs are ready to be harvested. The bulbs are dug up and allowed to dry for several weeks in a well ventilated area before they can be used or stored.

Ripeness To determine if a garlic bulb is ripe, check the outer layer of the bulb. It should be dry and papery, and the cloves should be plump and firm. If the outer layer is still moist or the cloves are shriveled or soft, the garlic may not be ripe or may be starting to spoil.

Spoilage Spoiled garlic will have a sour smell and may have mold or fungus growing on it. The cloves may also be soft or shriveled. If you notice any of these signs, discard the garlic to prevent the spread of bacteria.

Storing Garlic should be stored in a cool, dry place with good air circulation, away from direct sunlight. A dark pantry or a cabinet is a good spot. If you do not plan on using the garlic within a few weeks, you can store it in the refrigerator, but not for too long. Avoid storing garlic in plastic bags or containers, as this can promote moisture buildup and mold growth.

Preserving One simple way to preserve garlic is to roast it. Simply cut off the top of a bulb of garlic, drizzle with olive oil, wrap in foil, and bake in the oven for 30-40 minutes at 150°C. The roasted garlic can be used in a variety of dishes or spread on bread.

SAUTÉED GARLIC

Serves: 2
Prep Time: 5 minutes
Cook Time: 3 minutes

1 head of garlic
30 ml of extra virgin olive oil
Salt and pepper, to taste

Directions

1. Peel the garlic cloves and chop them finely.
2. Heat the olive oil in a skillet over medium heat.
3. Add the garlic and sauté for 1-2 minutes until fragrant.
4. Season with salt and pepper, to taste.

Notes
To add some extra flavor, consider adding a pinch of red pepper flakes or a splash of lemon juice.

ROASTED GARLIC OIL

Serves: 1
Prep Time: 10 minutes
Cook Time: 30 minutes

1 head of garlic
60 ml of extra virgin olive oil

Directions

1. Preheat oven to 150°C (300°F).
2. Peel the garlic cloves and cut them into long thin slices.
3. In a small oven safe glass dish, combine the garlic and olive oil.
4. Bake for 20-30 minutes or until the garlic is tender and lightly golden brown. Allow it to cool before using.
5. If desired, remove garlic pieces

Notes
Store the roasted garlic oil in an airtight container in the refrigerator for up to 1 week. Use it as a salad dressing or as a flavorful drizzle for any dish. Try substituting olive oil for any other desired oil.

AIR FRIED GARLIC

Serves: 1
Prep Time: 5 minutes
Cook Time: 10 minutes

1 head of garlic
30 ml of extra virgin olive oil
Salt and pepper, to taste

Directions

1. Preheat your air fryer to 200°C (400°F).
2. Peel the garlic cloves and chop them into small pieces.
3. Toss the garlic with olive oil and season with salt and pepper in a small oven safe glass dish.
4. Place the dish in the air fryer basket and cook for 8-10 minutes, occasionally stirring, until golden brown and crispy.

Notes
For a delicious twist, toss the cooked air-fried garlic with a blend of fresh herbs like parsley, basil, or cilantro. Enjoy it as a topping for salads, pasta, or as a garnish for your favorite dishes.

GINGER

History Ginger is a flowering plant that belongs to the Zingiberaceae family. It is native to Southeast Asia and has been used for its medicinal and culinary properties for over 5000 years. Ginger was highly valued in ancient India and China and was traded along the Silk Road, reaching the Middle East and Europe. During the Middle Ages, ginger was a popular spice in Europe and was used to flavor food and beverages, as well as for medicinal purposes.

Can you eat it raw? Yes, ginger can be eaten raw. It has a pungent, spicy, and slightly sweet taste, and is often used in Asian cuisine in its raw form. Raw ginger is also used to make ginger tea, which is believed to have several health benefits, including reducing inflammation and aiding digestion.

Growing and harvesting Ginger is a tropical plant that grows best in warm and humid climates. It is typically grown from rhizomes, which are underground stems. The plant grows to about 2-3 feet tall and produces green, lance shaped leaves and yellow flowers. Ginger is typically harvested 8-10 months after planting by pulling the roots from the soil.

Ripeness When selecting ginger, look for firm, smooth, and unblemished rhizomes with a fresh aroma. The skin of the rhizome should be thin and easy to peel. The flesh of the ginger should be pale yellow and juicy, with no signs of mold or spoilage.

Spoilage Spoiled ginger will have a soft, mushy texture and a sour smell. Mold may also appear on the surface of the rhizome. It's important to discard spoiled ginger, as it can cause foodborne illness.

Storing Fresh ginger can be stored in the refrigerator for up to three weeks. To store ginger for longer periods, it can be peeled, sliced, and stored in the freezer for up to six months. Alternatively, ginger can be stored in a cool, dry place in a paper bag or airtight container for up to two weeks.

Preserving One of the simplest ways to preserve ginger is to pickle it. To pickle ginger, slice it thinly and place it in a mixture of rice vinegar, sugar, and salt. Let it sit for a few hours before serving. Ginger can also be candied by simmering sliced ginger in a sugar syrup until it is translucent and then coating it in sugar. Candied ginger can be stored in an airtight container for several months.

GINGER STIR FRY

Serves 2
Prep Time: 10 minutes
Cook Time: 5 minutes

125 g ginger, peeled and julienned
30 ml vegetable or sesame oil
15 ml rice vinegar
Salt and pepper, to taste

Directions

1. Heat the vegetable oil in a wok or large skillet over high heat.
2. Add the ginger and stir fry for 2-3 minutes, or until fragrant.
3. Add the rice vinegar, salt, and pepper, and continue stir frying for another minute.
4. Remove from heat and serve hot as a side dish or over rice.

Notes

For added flavor, you can also stir fry other vegetables like bell peppers, carrots, or onions with the ginger.

GINGER SOUP

Serves: 4
Prep Time: 10 minutes
Cook Time: 30 minutes

450 g ginger, peeled and sliced
1 L water or stock
15 ml sesame oil
Salt and pepper, to taste

Directions

1. Heat the oil in a large pot over medium heat.
2. Add the ginger and sauté for 2-3 minutes.
3. Add the water and salt, and bring to a boil.
4. Reduce the heat, add the pepper, and simmer for 30 minutes.
5. Remove from heat and serve hot as a soup or strain and use as a base for other soups.

Notes

You can add other ingredients to the soup, such as chicken or vegetables, for more flavor and nutrition.

AIR FRIED GINGER FRIES

Serves: 1
Prep Time: 10 minutes
Cook Time: 12 minutes

150 g ginger, peeled and cut into thin fries
30 ml vegetable oil
Salt and pepper, to taste

Directions

1. Preheat the air fryer to 200°C (400°F).
2. Toss the ginger fries with oil, salt, and pepper.
3. Place the fries in the air fryer basket and cook for 10-12 minutes or until crispy.
4. Remove from the air fryer basket and serve hot as a snack or side dish.

Notes

You can also sprinkle sesame seeds or chili flakes on top of the ginger fries for added flavor and texture.

GREEN BEANS

History Green beans, or snap beans, are believed to have originated in ancient Mesoamerica, which is now modern-day Mexico and Central America, and were an important crop for the indigenous people of the region. They were introduced to Europe by Spanish explorers in the 16th century and were widely adopted as a food crop. Green beans were not commonly consumed in the United States until the 19th century, when French immigrants introduced them as a delicacy.

Can you eat it raw? Yes, green beans can be eaten raw, and are often used in salads or as a crunchy snack. However, some people find them tough and fibrous when eaten raw, and may prefer them cooked.

Growing and harvesting Green beans are a warm weather crop that prefer well draining soil and full sun exposure. They are typically planted in the spring and harvested in the summer, and can be grown on a trellis or left to sprawl on the ground. Green beans are harvested by hand or with a mechanical harvester once they reach maturity, which is typically when they are young and tender.

Ripeness Green beans should be picked when they are young and tender, which is when they are at their peak ripeness. They should be bright green, firm, and free of blemishes or spots. If the beans are dull or discolored, or if they feel limp or rubbery, they are past their prime and should be avoided.

Spoilage Spoiled green beans may have a sour smell, be slimy to the touch, or have black or brown spots. If the beans are discolored, soft, or have a bad odor, they should be discarded.

Storing Green beans should be stored in a plastic bag in the refrigerator for up to five days. To extend their shelf life, blanch the beans in boiling water for two to three minutes, then cool and store them in an airtight container in the freezer for up to six months.

Preserving Green beans can be canned or pickled to extend their shelf life. To can green beans, clean and trim them, then pack them into jars with a solution of water and vinegar. Process the jars in a boiling water bath for 20-30 minutes.

SAUTÉED GREEN BEANS

Serves: 4
Prep Time: 5 minutes
Cook Time: 7 minutes

450 g green beans, trimmed
30 ml extra virgin olive oil
Fresh lemon juice (optional)
Salt and pepper, to taste

Directions

1. In a large skillet, heat the olive oil over medium heat.
2. Add the green beans and sauté for 5-7 minutes, stirring occasionally, until they are crisp and tender.
3. Remove from heat and toss with lemon juice, salt, and pepper.
4. Serve warm.

Notes

Add toasted almond flakes or garlic for an extra layer of flavor.

STEAMED GREEN BEANS

Serves: 4
Prep Time: 5 minutes
Cook Time: 7 minutes

450 g green beans, trimmed
15 ml extra virgin olive oil
Salt and pepper, to taste

Directions

1. Fill a pot with 2.5 cm of water and bring to a boil.
2. Place a steamer basket in the pot, then add the green beans.
3. Cover the pot and steam for 5-7 minutes, until the beans are tender.
4. Remove from heat and toss with olive oil, salt, and pepper.
5. Serve warm.

Notes

Add a sprinkle of lemon zest or grated Parmesan cheese for a flavorful twist.

AIR FRIED GREEN BEANS

Serves: 2
Prep Time: 5 minutes
Cook Time: 10 minutes

225 g green beans, trimmed
45 ml extra virgin olive oil
Salt and pepper, to taste

Directions

1. Preheat the air fryer to 200°C (400°F).
2. Toss the green beans with olive oil, salt, and pepper.
3. Place the green beans in the air fryer basket and cook for 8-10 minutes, shaking the basket halfway through, until the beans are crispy and browned.
4. Remove from the air fryer and serve immediately.

Notes

Experiment with different seasonings, such as garlic powder or onion powder, for a unique flavor profile.

JICAMA

History Jicama, also known as yam bean, is a root vegetable that originated in Mexico and Central America. It is a member of the legume family and has been cultivated for thousands of years. The jicama plant was highly valued by the Aztecs and Mayas for its edible root, and it played an important role in their diets. Jicama was also used for medicinal purposes, as its juice was believed to have diuretic properties and could be used to treat coughs and colds.

Can you eat it raw? Yes, jicama can be eaten raw. In fact, it is often enjoyed in its raw state, either on its own as a snack or added to salads and slaws for a crunchy texture. The flesh of jicama has a crisp texture that is similar to water chestnuts, and its mild, sweet flavor makes it a versatile ingredient in both sweet and savory dishes.

Growing and harvesting Jicama is a vine that grows best in warm, tropical climates. It requires a long growing season of 9-10 months, making it a challenging crop to grow in some regions. The plant can grow up to 20 feet long, and its leaves are compound and pinnate, with leaflets that are 2-6 inches long. The root itself grows underground and is harvested when it reaches maturity, which is typically signaled by the leaves beginning to turn yellow.

Ripeness When selecting jicama at the grocery store or farmers' market, look for those that are firm and free of bruises or soft spots. The skin should be smooth and unblemished. Additionally, the larger the jicama, the more mature it is likely to be. To test for ripeness, give the jicama a firm squeeze. If it gives slightly under pressure but doesn't feel soft or mushy, it is likely ripe.

Spoilage Spoiled jicama will have a soft or mushy texture, and the skin may become discolored or shriveled. If the jicama emits an unpleasant odor, it is likely past its prime and should be discarded.

Storing Jicama can be stored in a cool, dry place for up to a month. Do not store it in the refrigerator, as temperatures below 50°F can damage the texture and flavor of the root. Once cut, wrap the unused portion tightly in plastic wrap and refrigerate for up to a week.

Preserving Jicama can be frozen for up to six months. To freeze, peel and cut into cubes, blanch for three minutes in boiling water, then plunge into ice water to stop the cooking process. Drain well and transfer to a freezer safe container. Frozen jicama may lose some of its crisp texture but is still suitable for use in soups, stews, and casseroles.

JICAMA SALAD

Serves: 2
Prep Time: 10 minutes
Cook Time: 5 minutes

1 medium jicama, peeled and julienned
15 ml extra virgin olive oil
15 ml fresh lime juice
Salt and pepper, to taste

Directions

1. In a large bowl, combine the jicama, olive oil, lime juice, salt, and pepper.
2. Toss to combine and coat the jicama evenly.
3. Let the salad rest for 5-15 minutes to allow the flavors to meld.
4. Serve immediately.

Notes

Add thinly sliced red onions, cilantro, and diced avocado for extra flavor and texture. Adjust salt and lime juice to taste.

JICAMA FRIES

Serves: 2
Prep Time: 10 minutes
Cook Time: 25 minutes

1 medium jicama, peeled and cut into fries
15 ml extra virgin olive oil
Salt and pepper, to taste

Directions

1. Preheat the oven to 200°C (400°F).
2. In a large bowl, toss the jicama fries with olive oil, salt, and pepper.
3. Spread the fries out in a single layer on a baking sheet.
4. Bake for 20-25 minutes, flipping once halfway through, until the fries are crispy and golden brown.
5. Serve immediately.

Notes

Experiment with different seasonings such as garlic powder, smoked paprika, or cajun spice. Serve with your favorite dipping sauce like ketchup, aioli, or spicy mayo.

AIR FRIED JICAMA CHIPS

Serves: 2
Prep Time: 10 minutes
Cook Time: 12 minutes

1 medium jicama, peeled and sliced into thin rounds
30 ml extra virgin olive oil
15 ml fresh lime juice
Salt and pepper, to taste

Directions

1. Preheat the air fryer to 200°C (400°F).
2. In a large bowl, toss the jicama rounds with olive oil, lime juice, salt, and pepper.
3. Arrange the jicama rounds in a single layer on aluminum foil in the air fryer basket.
4. Air fry for 8-12 minutes, flipping once halfway through, until the chips are golden brown and crispy.
5. Serve immediately.

Notes

Store leftover chips in an airtight container to maintain crispiness. Reheat by placing in air fryer.

KALE

History Kale has a long and interesting history that dates back to ancient times. While its exact origins are not known, kale is believed to have originated in the eastern Mediterranean and Asia Minor regions. Kale has been cultivated for over 2,000 years, and it was a popular vegetable in ancient Rome. In fact, it was so popular that the Roman philosopher Pliny the Elder wrote about kale in his natural history encyclopedia, praising its health benefits and versatility. During the Middle Ages, kale was one of the most common green vegetables in Europe.

Can you eat it raw? Yes, kale can be eaten raw and is often used in salads and smoothies. Raw kale can be slightly bitter and tough, so it's best to massage the leaves with a bit of olive oil or lemon juice to help soften them and enhance their flavor.

Growing and harvesting Kale is a hardy vegetable that grows best in cool weather and can withstand frost. It can be grown in most regions and is relatively easy to cultivate. Kale grows in a rosette of leaves that are harvested when they reach about 8-10 inches in length. The outer leaves are usually harvested first, leaving the inner leaves to continue growing. To keep your kale plants productive, be sure to harvest the leaves regularly and avoid letting them become too mature and tough.

Ripeness When buying kale, look for firm, crisp leaves that are a bright green color. Avoid leaves that are yellow, brown, or wilted. The stems should be firm and not too thick. If the stems are too thick, it may indicate that the kale is too mature and tough.

Spoilage Kale can spoil quickly if not stored properly. Look for signs of wilting, browning, or mold on the leaves. If you notice any of these signs, discard the kale. It is best to use kale as soon as possible after harvesting or purchasing it.

Storage To store kale, wrap it in a damp paper towel and place it in a plastic bag. Store it in the refrigerator for up to a week. Avoid washing kale before storing it, as excess moisture can cause it to spoil more quickly. If you need to wash your kale, be sure to dry it thoroughly before storing it.

Preserving Kale can be blanched and frozen for later use. To blanch kale, bring a pot of water to a boil and add the kale. Cook for 2-3 minutes, then remove from the water and place in a bowl of ice water to cool. Once cooled, drain the kale and place it in a freezer safe container. Frozen kale can be used in soups, stews, and casseroles.

SAUTÉED KALE

Serves: 4
Prep Time: 10 minutes
Cook Time: 7 minutes

1 bunch of kale, washed and chopped (approx. 200g)
30 ml extra virgin olive oil
15 ml fresh lemon juice
Salt and pepper, to taste

Directions

1. In a large skillet, heat the olive oil over medium-high heat.
2. Add the chopped kale to the skillet and cook for 3-5 minutes, stirring occasionally until it starts to wilt.
3. Add the lemon juice, salt, and pepper and continue cooking for an additional 1-2 minutes, until the kale is tender but still bright green.
4. Serve immediately.

Notes

For an extra burst of flavor, consider adding minced garlic or red pepper flakes to the skillet with the kale.

KALE CHIPS

Serves: 4
Prep Time: 10 minutes
Cook Time: 12 minutes

1 bunch of kale, washed and dried (approx. 200g)
15 ml extra virgin olive oil
Salt and pepper, to taste

Directions

1. Preheat the oven to 180°C (350°F).
2. Remove the stems from the kale leaves and tear the leaves into bite-sized pieces.
3. In a large bowl, toss the kale with olive oil, salt, and pepper, ensuring each piece is coated evenly.
4. Spread the kale out in a single layer on a baking sheet.
5. Bake for 10-12 minutes, or until the kale is crispy and slightly browned.
6. Let the chips cool for a few minutes before serving.

Notes

Try adding nutritional yeast or your favorite seasoning blend before baking.

AIR FRIED KALE

Serves: 4
Prep Time: 10 minutes
Cook Time: 7 minutes

1 bunch of kale, washed (approx. 200g)
30 ml extra virgin olive oil
Balsamic vinegar (optional)
Salt and pepper, to taste

Directions

1. Preheat the air fryer to 200°C (400°F).
2. In a large bowl, toss the kale with olive oil, balsamic, salt, and pepper, ensuring each piece is coated evenly
3. Place the kale on aluminum foil in the air fryer basket, spreading it out in a single layer and place stainless steel utensils over the kale to prevent it from moving while cooking.
4. Air fry for 5-7 minutes, until the kale is crispy and slightly browned.
5. Remove the kale from the air fryer and serve.

Notes

To add sweetness, drizzle with a bit of honey or maple syrup before serving.

KOHLRABI

History Kohlrabi, a cruciferous vegetable, has been cultivated in Europe since the 16th century. Its name comes from the German words "kohl" meaning cabbage and "rabi" meaning turnip, which is an apt description of its appearance.

Can you eat it raw? Kohlrabi can be consumed both raw and cooked. When eaten raw, it has a crisp texture and a mild, slightly sweet flavor that is comparable to an apple or a mild radish. When cooked, it has a tender texture and a slightly sweet, nutty flavor that pairs well with different seasonings and ingredients.

Growing and harvesting Kohlrabi is grown and harvested similarly to other cruciferous vegetables such as broccoli and cauliflower. It grows best in cooler temperatures and can be cultivated in the spring and fall in most regions. The vegetable is typically harvested when the bulb is about 2-3 inches in diameter, although it can be harvested at a larger size as well. When selecting kohlrabi, choose bulbs that are firm and free from bruises or blemishes.

Ripeness To determine if kohlrabi is ripe, gently press the bulb with your fingers. If it feels firm and solid, it is ready to eat. If it feels soft or spongy, it may be overripe or starting to spoil.

Spoilage Kohlrabi that is starting to spoil will feel soft and spongy, and it may also develop blemishes or mold. Avoid consuming spoiled kohlrabi as it can cause illness.

Storing To store kohlrabi, remove the leaves and keep the bulb in a plastic bag in the refrigerator. It can last up to a week. The leaves can be stored separately in a plastic bag in the refrigerator for up to three days.

Preserving One easy way to preserve kohlrabi is by blanching and freezing it. First, peel the kohlrabi and cut it into small pieces. Blanch the pieces in boiling water for 2-3 minutes, then transfer them to a bowl of ice water to cool. Drain the pieces and place them in a freezer safe container. They will keep in the freezer for up to six months.

KOHLRABI SALAD

Serves: 4
Prep Time: 10 minutes
Cook Time: 5

1 medium kohlrabi, peeled and thinly sliced
30 ml extra virgin olive oil
15 ml fresh lemon juice
Salt and pepper, to taste

Directions

1. In a large bowl, whisk together the olive oil and lemon juice.
2. Add the sliced kohlrabi and toss to coat in the dressing.
3. Season with salt and pepper, to taste.
4. Let the salad sit for 5 minutes to allow the flavors to meld.

Notes
Add thinly sliced red onion or cucumber for extra flavor and texture. Garnish with fresh herbs like parsley, dill, or mint.

ROASTED KOHLRABI

Serves: 2
Prep Time: 10 minutes
Cook Time: 25 minutes

1 medium kohlrabi, peeled and chopped into bite sized pieces
15 ml extra virgin olive oil
Salt and pepper, to taste

Directions

1. Preheat the oven to 200°C (400°F).
2. In a large bowl, toss the chopped kohlrabi with the olive oil and season with salt and pepper.
3. Spread the kohlrabi out in a single layer on a baking sheet.
4. Roast in the oven for 20-25 minutes, or until tender and golden brown.
5. Season with additional salt and pepper, to taste.
6. Serve hot.

Notes
Add garlic powder, paprika, or cayenne pepper for a spicy twist. Mix kohlrabi with other root vegetables like carrots, parsnips, or beets for variety.

AIR FRIED KOHLRABI FRIES

Serves: 3
Prep Time: 10 minutes
Cook Time: 15 minutes

1 medium kohlrabi, peeled and cut into French fry shaped pieces
30 ml extra virgin olive oil
Salt and pepper, to taste

Directions

1. Preheat your air fryer to 200°C (400°F).
2. In a large bowl, toss the kohlrabi fries with the olive oil and season with salt and pepper.
3. Place the kohlrabi fries on aluminum foil in the air fryer basket in a single layer.
4. Air fry for 12-15 minutes, shaking the basket occasionally, until the fries are crispy and golden brown.
5. Season with additional salt and pepper, to taste.
6. Serve hot.

Notes
Sprinkle with grated Parmesan cheese and chopped parsley for a delicious twist. Serve with your favorite dipping sauce, such as aioli, ketchup, or a spicy mayo.

LEEKS

History Leeks have a long history, believed to have originated in Central Asia and then spreading to the Mediterranean region. They have been cultivated for over 3,000 years, with evidence of leek cultivation found in ancient Egyptian tombs. The ancient Greeks and Romans also enjoyed leeks in their diet, and the latter are known to have made a dish called "porrophagus" that consisted of leeks, onions, and garlic cooked with a barley-mush base.

Can you eat it raw? While leeks are typically cooked, the tender white and light green parts can be sliced thinly and added to salads for a crunchy texture and mild onion flavor. However, the tougher green leaves are generally not eaten raw due to their fibrous texture.

Growing and harvesting Leeks are grown from seed and are typically started indoors in late winter or early spring. The seedlings are then transplanted to the garden in early spring or early summer. They prefer well-drained soil and full sun exposure, but can tolerate some shade. Leeks are ready for harvest when the white base is at least one inch in diameter, usually around late summer to early fall.

Ripeness When selecting leeks, look for those with firm, straight, and white bases. The leaves should be bright green and not wilted. Larger leeks are not necessarily better; in fact, smaller leeks are often more tender and flavorful.

Spoilage Spoiled leeks will have a soft, slimy texture and a foul odor. Discard any leeks that have brown spots, are slimy, or have mold on them. It is important to store leeks properly to avoid spoilage.

Storing Unwashed leeks can be stored in the refrigerator for up to two weeks. To store, wrap them in a damp paper towel and place them in a plastic bag. Alternatively, you can store them in a container filled with water, changing the water every few days. Do not store leeks near fruits, as the ethylene gas produced by the fruit can cause the leeks to spoil more quickly.

Preserving The simplest way to preserve leeks is by freezing them. To do this, blanch the leeks for 2-3 minutes, plunge them into ice water, drain, and then pack them into airtight containers.

BRAISED LEEKS

Serves: 1
Prep Time: 5 minutes
Cook Time: 20 minutes

1 large leeks, trimmed and halved lengthwise
15 ml extra virgin olive oil
30 ml water or stock
Balsamic vinegar (optional)
Salt and pepper, to taste

Directions

1. Heat the olive oil in a large skillet over medium heat.
2. Add the leeks cut side down and cook until browned, about 5 minutes.
3. Add the water, vinegar, salt, and pepper to the skillet and bring to a simmer.
4. Reduce the heat to low, cover, and cook until the leeks are tender, about 15 minutes.
5. Remove the leeks from the skillet and arrange them on a serving platter.
6. Drizzle the remaining cooking liquid over the leeks and serve hot.

Notes

For a richer flavor, consider adding a bouquet garni after lowering the heat to simmer.

GRILLED LEEKS

Serves: 1
Prep Time: 5 minutes
Cook Time: 10 minutes

1 large leeks, trimmed and cut into small pieces
15 ml extra virgin olive oil
Frsh lemon juice (optional)
Salt and pepper, to taste

Directions

1. Preheat grill to medium-high heat.
2. Toss the leeks with olive oil, lemon juice, salt, and pepper.
3. Grill the leeks until charred and tender, about 8-10 minutes, flipping once.
4. Remove the leeks from the grill and arrange them on a serving platter.
5. Drizzle any remaining marinade over the leeks and serve hot.

Notes

For added flavor, sprinkle the grilled leeks with fresh herbs like parsley or thyme before serving.

AIR FRIED LEEKS

Serves: 1
Prep Time: 5 minutes
Cook Time: 10 minutes

1 large leeks, trimmed and sliced into thin rounds
30 ml extra virgin olive oil
Salt and pepper, to taste

Directions

1. Preheat the air fryer to 200°C (400°F).
2. Toss the leeks with olive oil, salt, and pepper.
3. Spread the leeks on aluminum foil in a single layer in the air fryer basket.
4. Air fry for 5-10 minutes or until crispy and golden brown.
5. Remove the leeks from the air fryer and transfer them to a serving platter.
6. Serve hot.

Notes

For a spicy twist, add a pinch of paprika or cayenne pepper to the seasoning mix before air frying.

ROMAINE LETTUCE

History Romaine lettuce, also known as cos lettuce, is believed to have originated in the eastern Mediterranean over 5,000 years ago. It was first cultivated on the Greek island of Cos and was named after the island. The lettuce was introduced to ancient Rome and became popular in Europe during the 16th century.

Can you eat it raw? Yes, romaine lettuce is often eaten raw, as it is a common ingredient in salads, sandwiches, and wraps. However, it can also be cooked, grilled, or roasted.

Growing and harvesting Romaine lettuce is typically grown in rows on a farm. It requires full sun and fertile, well-drained soil to grow properly. The lettuce is usually planted in the spring and fall and takes about 60-70 days to mature. Romaine lettuce is harvested by cutting the entire head off at the base of the plant. The outer leaves are usually removed, leaving the inner leaves intact.

Ripeness When selecting romaine lettuce, look for crisp, green leaves that are tightly packed together. The leaves should not be wilted or have any brown spots. The outer leaves may be slightly darker in color than the inner leaves. The heads should feel firm when gently squeezed.

Spoilage Spoiled romaine lettuce will have wilted, slimy, or brown leaves. There may also be a foul odor coming from the lettuce. It is best to discard any lettuce that shows signs of spoilage.

Storing To store romaine lettuce, remove any damaged or wilted leaves and wrap the lettuce in a damp paper towel. Place the lettuce in a plastic bag and store it in the refrigerator's crisper drawer. Romaine lettuce will stay fresh for up to a week when stored properly.

Preserving Romaine lettuce can be preserved by blanching and freezing it. First, clean the lettuce thoroughly and cut it into bite sized pieces. Then, bring a pot of water to a boil and add the lettuce. Cook for 2-3 minutes, then transfer to an ice bath to stop the cooking process. Drain the lettuce and store it in an airtight container in the freezer for up to 6 months.

GRILLED ROMAINE LETTUCE

Serves: 2
Prep Time: 5 minutes
Cook Time: 3 minutes

1 head romaine lettuce, cleaned and dried
30 ml extra virgin olive oil
15 ml lemon juice or balsamic vinegar
Salt and pepper, to taste

Directions

1. Preheat a grill to medium-high heat.
2. Cut the romaine lettuce in half lengthwise, keeping the root end intact.
3. Brush the cut sides with olive oil and season with salt and pepper.
4. Place the lettuce cut side down on the grill and cook for 2-3 minutes or until grill marks appear.
5. Flip the lettuce and cook for an additional minute.
6. Remove from the grill and drizzle with lemon juice.
7. Serve immediately.

Notes
For an extra burst of flavor, sprinkle the grilled romaine lettuce with freshly grated Parmesan

ROMAINE LETTUCE SALAD

Serves: 2
Prep Time: 10 minutes
Cook Time: 0 minutes

1 head romaine lettuce, chopped
30 ml extra virgin olive oil
15 ml balsamic vinegar
Salt and pepper, to taste

Directions

1. In a large bowl, emulsify the olive oil and balsamic vinegar.
2. Season with salt and pepper, to taste.
3. Toss the salad to combine all the ingredients.
4. Serve immediately.

Notes
Enhance the salad by adding cherry tomatoes, cucumber, and croutons. You may also include crumbled feta or goat cheese for extra richness.

AIR FRIED ROMAINE LETTUCE

Serves: 2
Prep Time: 5 minutes
Cook Time: 5 minutes

1 head romaine lettuce, cleaned and dried
45 ml extra virgin olive oil
Salt and pepper, to taste

Directions

1. Preheat an air fryer to 220°C (430°F).
2. Quarter the romaine lettuce lengthwise, keeping the root end intact.
3. In a small bowl, comombine the olive oil and season with salt and pepper.
4. Brush the romaine quarters with the oil mixture
5. Place the lettuce on aluminum foil in the air fryer basket and cook for 5 minutes, or until the leaves are crispy and slightly browned.
6. Serve immediately.

Notes
To add some variety to this dish, sprinkle air fried romaine lettuce with garlic powder or onion powder before cooking. You may also add a drizzle of balsamic glaze on top before serving for a tangy twist.

MUSHROOM

History Mushrooms have a rich history that spans thousands of years. Evidence of mushroom consumption dates back to ancient civilizations such as Egypt and Greece, where they were considered a delicacy and believed to have healing properties. In ancient Rome, mushrooms were served as a dish of the elite, and in Asia, they were used for medicinal purposes. In the Middle Ages, mushrooms were highly prized and often reserved for the nobility.

Can you eat it raw? While mushrooms can be eaten raw, they are more commonly cooked. Raw mushrooms can be tough and difficult to digest for some people. It is recommended to cook mushrooms before consuming them to improve their texture and make them more palatable. Cooking mushrooms also enhances their flavor and aroma. Not all mushrooms are edible and can even be **poisonous.**

Growing and harvesting Mushrooms are grown in a controlled environment called a mushroom house or farm. The process of growing mushrooms is called cultivation, and it involves planting mushroom spores in a substrate, such as straw, sawdust, or compost. The substrate is kept in a dark, moist environment where the mushrooms can grow. When the mushrooms are mature, they are harvested by hand to ensure their quality.

Ripeness Mushrooms are at their best when they are firm, plump, and have a fresh aroma. The caps should be tightly closed, and the gills should be light in color. Mushrooms that are past their prime will have a slimy texture, a foul odor, and discoloration.

Spoilage Spoiled mushrooms will have a slimy texture, a strong odor, and discoloration. If you notice any of these signs, it is best to discard the mushrooms. Eating spoiled mushrooms can cause food poisoning and other health problems.

Storing To keep mushrooms fresh, they should be stored in the refrigerator in a paper bag or a container with a damp paper towel. Do not store them in a plastic bag or they may become slimy. Mushrooms should be used within a few days of purchase, and they can last up to five days when stored properly.

Preserving Mushrooms can be preserved by drying them. To do this, clean and slice the mushrooms and place them on a baking sheet in a single layer. Bake in a low oven until the mushrooms are dry and crispy. Store in an airtight container in a cool, dry place. Dried mushrooms can be rehydrated and used in a variety of dishes, such as soups, stews, and sauces.

SAUTÉED MUSHROOM

Serves: 4
Prep Time: 10 minutes
Cook Time: 7 minutes

450 g mushrooms, cleaned and sliced
15 ml extra virgin olive oil
Fresh lemon juice (optional)
Salt and pepper, to taste

Directions

1. Clean the mushrooms and slice them.
2. Heat the oil in a skillet over medium heat.
3. Add the mushrooms and sauté for 5-7 minutes, or until they are golden brown and tender.
4. Add the acid and season with salt and pepper, to taste.
5. Serve hot.

Notes

For added flavor, consider adding minced garlic or fresh herbs such as thyme or parsley during the sautéing process.

MUSHROOM SOUP

Serves: 4
Prep Time: 10 minutes
Cook Time: 20 minutes

450 g mushrooms
30 ml extra virgin olive oil
500 ml water or stock
Salt and pepper, to taste

Directions

1. Clean the mushrooms and slice them.
2. Heat the oil in a pot over medium heat.
3. Add the mushrooms and sauté for 5-7 minutes, or until they are golden brown and tender.
4. Add the water and bring to a boil.
5. Reduce heat and simmer for 10-15 minutes.
6. Season with salt and pepper, to taste.
7. Serve hot.

Notes

To make creamier, add heavy cream or coconut milk after simmering and blend the soup with an immersion blender until smooth.

AIR FRIED PORTOBELLO MUSHROOM CAPS

Serves: 1
Prep Time: 5 minutes
Cook Time: 18 minutes

1 large Portobello mushroom caps, cleaned and dried
15 ml extra virgin olive oil
Salt and pepper, to taste

Directions

1. Clean the mushrooms and pat them dry.
2. Preheat the air fryer to 220°C (430°F).
3. Coat the entire mushroom cap with oil, salt, and pepper.
4. Place the mushrooms in the air fryer basket and cook for 10-18 minutes, or until they are crispy.
5. Serve hot.

Notes

To enhance the flavor, sprinkle with garlic powder, onion powder, or your favorite seasoning blend before cooking.

OKRA

History Okra is a vegetable that originated in Ethiopia, and then spread to other parts of Africa before being introduced to the rest of the world through the slave trade. It was first brought to the Americas in the 17th century and is now widely cultivated in many parts of the world, including Asia and the Americas. Okra is a popular ingredient in Southern cuisine in the United States, particularly in dishes such as gumbo.

Can you eat the vegetable raw? Yes, you can eat okra raw, and it has a slightly sweet and grassy flavor. Raw okra is commonly used in salads, as a snack or as a garnish.

Growing and harvesting Okra plants are warm season crops that thrive in tropical and subtropical climates. They prefer well drained soil with a pH of 6.0 to 6.8 and full sun. Okra can be started from seed indoors 4-6 weeks before the last frost or sown directly into the garden after the last frost. The plants should be spaced about 12-18 inches apart. Okra pods are typically harvested when they are 2-4 inches long, still tender, and snap easily when bent.

Ripeness Okra should be harvested when the pods are 2-4 inches long, still tender, and snap easily when bent. If the pods are left on the plant for too long, they can become tough and fibrous.

Spoilage Spoiled okra will have a slimy texture, off odor, and discoloration. It is important to inspect okra before purchasing, and store it properly to prevent spoilage.

Storing Okra can be stored in a perforated plastic bag in the refrigerator for up to 5 days. It is important to store okra in a dry environment to prevent spoilage. Avoid washing okra before storing as moisture can cause it to spoil quickly.

Preserving Okra can be preserved by blanching and freezing. To blanch okra, bring a pot of water to a boil and add the okra pods. Cook for 3-4 minutes, then remove from the water and place in an ice bath to cool. Once cool, drain the okra and store in an airtight container in the freezer for up to 8 months.

ROASTED OKRA

Serves: 4
Prep Time: 10 minutes
Cook Time: 25 minutes

225 g fresh okra
15 ml extra virgin olive oil
Balsamic glaze (optional)
Salt and pepper, to taste

Directions

1. Preheat the oven to 220°C (430°F).
2. Trim the stems from the okra pods and slice them in half lengthwise.
3. In a large bowl, toss the okra with olive oil, balsamic, salt, and pepper.
4. Spread the okra out in a single layer on a baking sheet.
5. Roast in the preheated oven for 20-25 minutes, or until tender and lightly browned.
6. Serve hot.

Notes

For a variation, sprinkle some grated Parmesan cheese on top of the roasted okra before serving.

OKRA SALAD

Serves: 2
Prep Time: 15 minutes
Cook Time: 0

225 g fresh okra
30 ml extra virgin olive oil
30 ml fresh lemon juice
Salt and pepper, to taste

Directions

1. Trim the stems from the okra pods and slice them in half lengthwise.
2. In a large bowl, whisk together the olive oil, lemon juice, salt, and pepper.
3. Add the okra to the bowl and toss to coat with the dressing.
4. Serve immediately or refrigerate until ready to serve.

Notes

This salad can also be served with tomatoes and crumbled feta cheese on top for added flavor.

AIR FRIED OKRA

Serves: 4
Prep Time: 10 minutes
Cook Time: 10 minutes

225 g fresh okra
30 ml vegetable oil
Fresh lemon juice (optional)
Salt and pepper, to taste

Directions

1. Trim the stems from the okra pods and slice them in half lengthwise.
2. In a large bowl, toss the okra with olive oil, lemon juice, salt, and pepper.
3. Preheat the air fryer to 220°C (430°F).
4. Place the okra on aluminum foil in the air fryer basket in a single layer.
5. Cook for 8-10 minutes, shaking the basket occasionally, or until the okra is tender and lightly browned.
6. Serve hot.

Notes

For a variation, sprinkle some Cajun seasoning on the okra before air frying for added flavor.

ONION

History Onions have a rich history that spans thousands of years. They are believed to have been domesticated in central Asia and have been cultivated for over 5,000 years. Onions were highly valued in ancient civilizations, such as the Egyptians, who used them as currency, and the Greeks and Romans, who believed in their medicinal properties. Onions also played a significant role in the diet of many cultures and were used for cooking and as a staple food source during times of scarcity.

Can you eat it raw? Onions can be eaten raw, although their pungent flavor is often more palatable when cooked. Raw onions are commonly used in salads or as a garnish, but they can also be pickled or added to sandwiches and wraps.

Growing and harvesting Onions prefer well draining soil and a location with full sun exposure. They can be grown from seeds or sets, which are small onion bulbs. Onions are typically ready to harvest when the leaves turn yellow and begin to dry out. Once harvested, onions should be cured by drying them in a warm, dry place for several weeks to improve their flavor and shelf life.

Ripeness Ripe onions should be firm with a dry outer skin. The neck should be tight and not show any signs of sprouting. The size of the onion bulb will vary depending on the variety and growing conditions.

Spoilage Spoiled onions may have soft spots or black mold on the outside. They may also have a sour or musty odor. It is important to inspect onions for any signs of spoilage before using or storing them.

Storing Onions should be stored in a cool, dry place with good ventilation. Avoid storing onions with potatoes, as they can release gases that cause each other to spoil more quickly. Onions can be stored for several weeks to several months, depending on the variety.

Preservating Onions can be preserved in a variety of ways like freezing or dehydrating. The simplest way to freeze onions is to chop or slice them and store them in an airtight container in the freezer. Onions can also be pickled, which extends their shelf life and adds a tangy flavor to dishes.

CARAMELIZED ONIONS

Serves: 3
Prep Time: 5 minutes
Cook Time: 1-8 hours

4 large onions, sliced or chopped
15 ml extra virgin olive oil
Salt and pepper, to taste

Directions

1. In a large skillet, heat oil over low-medium heat.
2. Add onions and stir often enough that the onions do not burn while cooking.
3. Cook onions until desired doneness.

Notes
The lower the heat and the slower you cook them, the more flavorful they will be.

ROASTED ONION

Serves: 2
Prep Time: 5 minutes
Cook Time: 25 minutes

1 large onion, sliced into rounds
15 ml extra virgin olive oil
Balsamic glaze (optional)
Salt and pepper, to taste

Directions

1. Preheat oven to 200°C (400°F).
2. Toss sliced onions with oil and balsamic and season with salt and pepper.
3. Arrange onions on a baking sheet and roast for 20-25 minutes until golden brown and tender.

Notes
Make sure to spread the onions out in a single layer on the baking sheet so that they roast evenly. Using balsamic adds a slight sweetness and tang to the onions, but you can also use other types of vinegar if you prefer.

AIR FRIED ONION RINGS

Serves: 2
Prep Time: 10 minutes
Cook Time: 12 minutes

1 large onion, sliced into rings
45 ml vegetable oil
15 ml fresh lemon juice (optional)
Salt and pepper, to taste

Directions

1. Preheat air fryer to 200°C (400°F).
2. Toss onion rings with oil and season with salt and pepper.
3. Arrange onion rings on aluminum foil in a single layer in the air fryer basket and cook for 10-12 minutes, flipping halfway through, until golden brown and crispy.

Notes
To have a classic battered onion ring, coat the onion ring in an egg mixture and dredge in seasoned flour mixture. You can also add a pinch of cayenne pepper or paprika for some extra flavor.

PARSNIP

History The history of parsnips is not well documented, but they are believed to have originated in Eurasia and have been used in European cuisine for centuries. They were also a staple food in ancient Rome and Greece, and were likely brought to North America by early European settlers.

Can you eat it raw? Parsnips are typically cooked before eating, as they can be tough and woody when raw. However, some people do enjoy eating parsnips raw, either grated into salads or sliced thinly and served with dips.

Growing and harvesting Parsnips are cool season vegetables that prefer well draining soil and full sun. They can be grown from seed sown in early spring, and take 5-6 months to mature. It's important to keep the soil evenly moist during germination, and to thin the seedlings to prevent overcrowding. When the leaves of the plant begin to yellow and die back, it is time to harvest the parsnips. They should be dug up carefully to avoid damaging the roots.

Ripeness Parsnips are ready to harvest when they are about 8-10 inches long and 1-2 inches in diameter. The skin should be smooth and free from blemishes. The flesh should be firm and not too woody. Some gardeners prefer to leave parsnips in the ground until after a frost, as the cold temperature can enhance their flavor.

Spoilage Spoiled parsnips will have soft spots, discolored skin, and an unpleasant odor. They should be discarded immediately. To prevent spoilage, it's important to store parsnips properly.

Storing Parsnips can be stored in a cool, dark, and humid place for up to 2-3 weeks. They should be stored in a perforated plastic bag or wrapped in damp paper towels to prevent them from drying out. Alternatively, they can be stored in a root cellar or refrigerator, where the temperature is around 32-40°F.

Preserving Parsnips can be preserved by freezing. To freeze, blanch the parsnips for 2-3 minutes, then shock them in cold water. Pat them dry and pack them into freezer safe containers.

ROASTED PARSNIPS

Serves: 2
Prep Time: 10 minutes
Cook Time: 30 minutes

225 g parsnips, peeled and chopped into bite sized pieces
30 ml extra virgin olive oil
Balsamic glaze (optional)
Salt and pepper, to taste

Directions

1. Preheat the oven to 200°C (400°F).
2. In a large mixing bowl, toss the parsnips with olive oil, balsamic, salt, and pepper.
3. Spread the parsnips in a single layer on a baking sheet.
4. Roast for 25-30 minutes, or until the parsnips are tender and golden brown.
5. Remove from the oven and serve hot.

Notes

For additional flavor, consider adding herbs such as rosemary or thyme to the parsnips before roasting.

BOILED PARSNIPS

Serves: 2
Prep Time: 10 minutes
Cook Time: 20 minutes

225 g parsnips, peeled and sliced
500 ml water or stock
Balsamic glaze (optional)
Salt and pepper, to taste

Directions

1. In a pot, bring the water to a boil.
2. Add the parsnips and lemon juice to the pot.
3. Reduce the heat to medium-low and simmer for 15-20 minutes, or until the parsnips are tender.
4. Drain the parsnips and season with salt and pepper.
5. Transfer the boiled parsnips to a serving dish and serve hot.

Notes

For a creamier texture, consider mashing the boiled parsnips with some butter and cream

AIR FRIED PARSNIP FRIES

Serves: 2
Prep Time: 10 minutes
Cook Time: 20 minutes

225 g parsnips, peeled and cut into fries
30 ml extra virgin olive oil
Salt and pepper, to taste

Directions

1. Preheat the ar fryer to 200°C (400°F).
2. In a large mixing bowl, toss the parsnips with the olive oil, salt, and pepper.
3. Place the parsnips on aluminum foil in the air fryer basket.
4. Air fry for 15-20 minutes, or until the parsnips are golden brown and crispy.
5. Remove from the air fryer and serve hot as a side dish or a snack.

Notes

For a spicier kick, consider adding cayenne pepper or chili powder to the parsnip fries before air frying.

PEAS

History Peas are one of the oldest cultivated vegetables in the world, dating back to at least 7,000 years ago. The exact origin of peas is unknown, but it is believed that they were first cultivated in the Mediterranean and Middle East regions. Ancient civilizations, including Greeks and Romans, considered peas a valuable food source and consumed them in various dishes. During the Middle Ages, peas were introduced to Europe and quickly became a popular crop.

Can you eat it raw? Yes, raw peas have a sweet taste and a slightly crunchy texture when eaten raw and are delicious and nutricious.

Growing and harvesting Peas are cool-season crops that thrive in temperatures ranging from 60-75°F (15-23°C). They grow best in well drained soil and require plenty of sunlight but can tolerate some shade. Depending on the variety, peas can be sown in either spring or fall. Peas are typically ready for harvest when the pods are plump and firm, but before the peas inside have become too large. Snow peas and snap peas are usually harvested when the pods are flat and the peas inside are still small.

Ripeness When selecting fresh peas, look for pods that are bright green and plump, and the peas inside should be small and firm. Avoid pods that are yellow or brown as they are overripe and may have a mealy texture.

Spoilage Peas can spoil quickly if they are not stored properly. Signs of spoilage include mold, discoloration, or a slimy texture. If you notice any of these signs, discard the peas.

Storing Fresh peas should be stored in the refrigerator in a plastic bag or container and should be used within five days. If you wish to store them for a longer period, blanch them for one to two minutes, cool, drain, and then store them in an airtight container or freezer bag in the freezer. Frozen peas can last up to eight months.

Preservating Freezing is one of the simplest and most effective ways to preserve peas. To do this, blanch the peas for one to two minutes, immediately transfer them to an ice water bath to stop the cooking process, and then drain them. Pack the peas into airtight containers or freezer bags and store them in the freezer. Frozen peas can be used in a variety of recipes, including soups, stews, and stir fries.

PEA SALAD

Serves: 3
Prep Time: 10 minutes
Cook Time: 2 minutes

200 g fresh peas
15 ml extra virgin olive oil
30 ml lemon juice or balsamic reduction
Salt and pepper, to taste

Directions:

1. Bring a large pot of salted water to a boil. Add the peas and blanch for 1-2 minutes, or until tender.
2. Drain the peas and transfer them to an ice water bath to stop the cooking process and cool them down.
3. In a large mixing bowl, whisk together the olive oil, lemon juice, salt, and pepper.
4. Add the peas to the bowl and toss to coat evenly.
5. Cover the bowl with plastic wrap and chill in the refrigerator for 30 minutes before serving.

Notes
For added flavor, you can include mint leaves, crumbled feta cheese, or chopped almonds to the salad.

ROASTED PEAS

Serves: 2
Prep Time: 5 minutes
Cook Time: 25 minutes

200 g fresh peas
30 ml extra virgin olive oil
Salt and pepper, to taste

Directions

1. Preheat the oven to 200°C (400°F).
2. Rinse and pat dry the peas with a paper towel. In a mixing bowl, toss the peas with olive oil, salt, and pepper until evenly coated.
3. Spread the peas in a single layer on a baking sheet.
4. Roast in the preheated oven for 20-25 minutes, stirring occasionally, until the peas are golden brown and crispy.
5. Remove from the oven and let them cool for a few minutes before serving.

Notes
For a spicy twist, you can sprinkle some chili flakes or cayenne pepper on the peas before roasting.

AIR FRIED PEAS

Serves: 2
Prep Time: 5 minutes
Cook Time: 10 minutes

200 g fresh peas
45 ml extra virgin olive oil
Salt and pepper, to taste

Directions

1. Rinse and pat dry the peas with a paper towel. In a medium sized oven-safe dish, toss the peas with olive oil, salt, and pepper until evenly coated.
2. Place the dish in the air fryer and set it to 200°C (400°F).
3. Air fry for 6-10 minutes, stirring every 2 minutes, until the peas are golden and crispy.
4. Remove from the air fryer and let them cool for a few minutes before serving.

Notes
You can experiment with different seasonings like garlic powder, onion powder, or paprika to customize the flavor of the air fried peas.

POTATO

History Potatoes are one of the most widely consumed vegetables in the world, and they have a rich history that dates back thousands of years. Potatoes are native to the Andean region of South America, where they were first cultivated over 7,000 years ago by indigenous communities. The Inca Empire, which flourished in the Andean region from the 13th to the 16th century, relied heavily on potatoes as a staple food crop.

Can you eat it raw? While potatoes are most commonly cooked, they can be eaten raw. However, it is important to note that raw potatoes contain solanine, a toxic compound that can cause digestive issues if consumed in large quantities. If you do choose to eat a raw potato, make sure to peel it first and only eat small amounts.

Growing and harvesting Potatoes are typically grown from "seed potatoes," which are small potatoes that have sprouted. They can be grown in a variety of climates and soils, but they prefer loose, well drained soil and cool temperatures. The planting season depends on the climate, but it is typically in the spring. The plant produces edible tubers underground, and they are usually harvested when the leaves of the plant start to turn yellow and die back, indicating that the potatoes are mature.

Ripeness When selecting potatoes, look for ones that are firm and free from blemishes.

Spoilage Potatoes can spoil relatively quickly if not stored properly. Look for signs of rotting, such as a foul smell, discoloration, or soft spots. If you notice any of these signs, discard the potato.

Storing Potatoes should be stored in a cool, dry, and dark place, such as a pantry or cellar. Exposure to light can cause potatoes to turn green and produce solanine. Potatoes should not be stored in the refrigerator, as this can cause them to develop a sweet flavor and turn black when cooked.

Preserving The simplest way to preserve potatoes is by storing them in a cool, dry place. If you have an abundance of potatoes, you can also can them or freeze them for later use. However, freezing may affect the texture and flavor of the potatoes. Blanching potatoes before freezing can help to maintain their texture and color.

ROASTED POTATOES

Serves: 4
Prep Time: 10 minutes
Cook Time: 30 minutes

4 medium-sized potatoes, cut into small pieces
30 ml butter or extra virgin olive oil
Balsamic vinegar (optional)
Salt and pepper, to taste

Directions

1. Preheat the oven to 200°C (400°F).
2. In a large bowl, toss the potatoes with olive oil and balsamic until evenly coated.
3. Season with salt and pepper according to your preference.
4. Spread the potatoes in a single layer on a baking sheet.
5. Roast for 25-30 minutes, or until the potatoes are crispy and golden brown, turning them halfway through the cooking time.

Notes

For a different flavor, try adding your favorite herbs or spices, such as rosemary, thyme, or garlic powder, to the oil and vinegar mixture before tossing the potatoes.

MASHED POTATOES

Serves: 2
Prep Time: 10 minutes
Cook Time: 20 minutes

4 medium-sized gold potatoes, quarted
30 g butter
Fresh lemon juice (optional)
Salt and white pepper, to taste

Directions

1. Place the potatoes in a medium sized pot and cover them with cold water. Add 1 pinch of salt per potato to the water.
2. Bring the water to a boil over high heat, then reduce the heat to medium-low and simmer until the potatoes are fork tender, about 15-20 minutes.
3. Drain the water from the pot.
4. Add the butter and lemon juice to the pot and toss the potatoes to coat.
5. Mash the potatoes to your desired consistency using a wooden spoon or mixer.
6. Season with salt and white pepper.

Notes

For added flavor, consider using stock instead of water or adding some minced garlic or chives to the mashed potatoes.

AIR FRIED POTATO FRIES

Serves: 2
Prep Time: 10 minutes
Cook Time: 15 minutes

3 medium-sized potatoes, cut into fries
60 ml oil or rendered fat
Salt and pepper, to taste

Directions

1. Preheat the air fryer to 220°C (430°F).
2. In a large bowl, toss the potatoes in oil or fat until evenly coated.
3. Transfer the potatoes to the air fryer basket on aluminum foil in a single layer.
4. Cook for 12-15 minutes, or until the potatoes are golden brown and crispy, shaking the basket halfway through the cooking time.

Notes

Cooking times may vary depending on the size and power of your air fryer. Adjust cooking time accordingly. For a different flavor, try seasoning the fries with your favorite spices, such as paprika or garlic powder, before air frying. The oil or fat used will impact the flavor of your fries.

PUMPKIN

History Pumpkins are a type of winter squash and are one of the oldest domesticated plants in North America, with evidence suggesting they were grown as early as 7,500 to 5,000 BCE by indigenous peoples. The name "pumpkin" comes from the Greek word "pepon," meaning "large melon." Pumpkins were initially cultivated for their seeds, which were an important source of protein for native peoples. Over time, selective breeding led to the development of pumpkins with larger, more fleshy fruit, which became a staple in the diet of many indigenous groups.

Can you eat it raw? While it is possible to eat pumpkin raw, it is not very common due to its tough, fibrous texture. It is more commonly cooked in a variety of ways, such as roasting, steaming, or boiling, which can help to soften the flesh and make it more palatable.

Growing and harvesting Pumpkins are typically grown on a vine and require full sun and well drained soil to thrive. They are typically planted in late spring or early summer and are ready to harvest in the fall, usually around September or October. When the stem connecting the pumpkin to the vine starts to dry and crack, the pumpkin is ready to be harvested.

Ripeness To determine if a pumpkin is ripe, it should be firm and heavy for its size. The skin should be a consistent, deep orange color, and there should be no soft spots or cracks. The stem should be hard and firmly attached to the pumpkin.

Spoilage A spoiled pumpkin will have soft spots, mold, or an off smell. If the skin is mushy or the pumpkin feels lighter than it should, it has likely gone bad and should be discarded.

Storing Whole pumpkins can be stored in a cool, dry place, such as a pantry or root cellar, for several months. Cut pumpkin should be wrapped in plastic and refrigerated for up to five days. Cooked pumpkin can also be frozen for later use.

Preserving The simplest way to preserve pumpkin is to roast it and puree it. The puree can then be stored in the freezer for several months or canned using a pressure canner. This can be a great way to enjoy pumpkin all year round, especially in pies, soups, and other dishes.

ROASTED PUMPKIN SALAD

Serves: 4
Prep Time: 10 minutes
Cook Time: 25 minutes

1 small pumpkin, peeled and chopped into pieces
30 ml extra virgin olive oil
Balsamic glaze (optional)
Salt and pepper, to taste
Mixed greens, for serving

Directions

1. Preheat the oven to 200°C (400°F).
2. In a bowl, toss the pumpkin pieces with olive oil, salt, and pepper.
3. Spread the pumpkin evenly on a baking sheet.
4. Roast for 20-25 minutes, or until tender and lightly browned.
5. Arrange the mixed greens on a platter and top with the roasted pumpkin.
6. Drizzle with additional olive oil and balsamic, if desired.

Notes
Try adding toasted nuts or seeds for extra crunch and flavor. Crumbled feta cheese or goat cheese would make a delicious addition.

STEAMED PUMPKIN

Serves: 2
Prep Time: 5 minutes
Cook Time: 12 minutes

1 small pumpkin, peeled and sliced into wedges
30 g butter
Fresh lemon juice (optional)
Salt and pepper, to taste

Directions

1. Place the pumpkin wedges in a steamer basket and steam for 10-12 minutes, or until tender.
2. In a small saucepan over low heat, melt the butter.
3. Stir in the lemon juice, salt, and pepper.
4. Drizzle the lemon butter over the steamed pumpkin and serve immediately.

Notes
Add some minced garlic or fresh herbs to the lemon butter for more flavor. Steamed pumpkin can also be served with a yogurt or sour cream based sauce.

AIR FRIED PUMPKIN FRIES

Serves: 2
Prep Time: 10 minutes
Cook Time: 20 minutes

1 small pumpkin, peeled and sliced into thin fries
60 ml extra virgin olive oil
Salt and pepper, to taste

Directions

1. Preheat the air fryer to 220°C (430°F).
2. In a bowl, toss the pumpkin fries with olive oil, salt, and pepper.
3. Place the pumpkin fries on aluminum foil in the air fryer basket in a single layer.
4. Air fry for 10-20 minutes, or until golden and crispy.
5. Serve immediately.

Notes
Experiment with different seasonings, such as garlic powder, onion powder, or smoked paprika, for a unique flavor twist. Serve with a dipping sauce, like aioli or yogurt-based sauce, for added enjoyment.

RADICCHIO

History Radicchio is a leafy vegetable that is a member of the chicory family, which also includes endive and escarole. It is believed to have originated in the Mediterranean region and has been cultivated in Italy for centuries, where it is a popular ingredient in Italian cuisine. It was first introduced to the United States in the late 19th century, but it was not until the late 20th century that it gained popularity in American cuisine.

Can you eat it raw? Yes, radicchio can be eaten raw. It is often used in salads and pairs well with a variety of other ingredients. Its slightly bitter taste can be balanced with sweeter flavors like citrus fruits or roasted vegetables.

Growing and harvesting Radicchio is typically grown in the cooler months of the year, as it prefers temperatures between 45 and 65 degrees Fahrenheit. It can be grown from seeds or transplanted seedlings and requires well draining soil with consistent moisture. It is usually planted in the fall and harvested in the winter or early spring. The leaves of the radicchio are harvested by cutting them off at the base of the plant. Some varieties of radicchio, such as the popular Chioggia variety, require a longer growing season and are usually harvested in the late spring or early summer.

Ripeness To tell if radicchio is ripe, it should be firm and brightly colored. The leaves should be crisp and free of any yellow or brown spots. The outer leaves of the radicchio can be slightly tougher and more bitter than the inner leaves.

Spoilage When radicchio has spoiled, it will be wilted and have a slimy texture. It may also have a sour or unpleasant smell. It is important to discard any radicchio that has spoiled as it may contain harmful bacteria.

Storing To store radicchio, remove any wilted or damaged leaves and wrap the remaining leaves in damp paper towels. Place the wrapped radicchio in a plastic bag and store it in the refrigerator. It should last for up to a week when stored in this way. Avoid washing the radicchio before storing it as excess moisture can lead to spoilage.

Preserving The simplest way to preserve radicchio is to blanch the leaves in boiling water for 1-2 minutes and then freeze them in an airtight container. This will allow you to use the radicchio later in soups or stews.

GRILLED RADICCHIO SALAD

Serves: 2
Prep Time: 5 minutes
Cook Time: 6 minutes

1 head of radicchio
45 ml extra virgin olive oil
15 ml balsamic reduction
Salt and pepper, to taste

Directions

1. Preheat a grill or grill pan to medium-high heat.
2. Cut the radicchio head into quarters, ensuring the core remains attached to hold the leaves together.
3. Brush the radicchio quarters with olive oil and season with salt and pepper.
4. Grill the radicchio quarters for 2 to 3 minutes per side, until lightly charred and softened.
5. Remove the radicchio from the grill and let cool for a few minutes.
6. Drizzle the balsamic vinegar over the radicchio and season with additional salt and pepper, if needed.

Notes

For additional flavor, consider adding crumbled goat cheese or shaved Parmesan and toasted pine nuts on top of the salad.

RADICCHIO SALAD

Serves: 2
Prep Time: 10 minutes
Cook Time: 0 minutes

1 head of radicchio
1 orange, peeled and supremed
30 ml extra virgin olive oil
30 ml fresh lemon juice
Salt and pepper, to taste

Directions

1. Cut the radicchio head in half and remove the core.
2. Slice the radicchio into very thin ribbons and transfer to a large bowl.
3. Add the orange segments to the bowl with the radicchio.
4. In a small bowl, whisk together the olive oil and lemon juice.
5. Pour the dressing over the radicchio and oranges and toss to combine.
6. Season with salt and pepper, to taste.

Notes

For added texture, consider including chopped walnuts or sliced almonds to the salad. You can also add pickled red onion for a burst of flavor.

AIR FRIED RADICCHIO CHIPS

Serves: 4
Prep Time: 5 minutes
Cook Time: 7 minutes

1 head of radicchio
30 ml extra virgin olive oil
Salt and pepper, to taste

Directions

1. Preheat an air fryer to 190°C (375°F).
2. Cut the radicchio head into bite sized pieces.
3. In a medium bowl, toss the radicchio with olive oil, salt, and pepper until evenly coated.
4. Place the radicchio on aluminum foil in the air fryer basket in a single layer.
5. Air fry the radicchio for 5 to 7 minutes, shaking the basket once or twice during cooking, until the leaves are crispy and lightly browned.
6. Transfer the radicchio chips to a serving platter and serve immediately.

Notes

You can experiment with different seasoning blends to add extra flavor to the radicchio chips. For example, try adding garlic powder, smoked paprika, or grated Parmesan cheese.

RADISH

History Radishes have a long and rich history that can be traced back to ancient times. It is believed that radishes were first domesticated in Southeast Asia, possibly in China or Japan, over 2,000 years ago. From there, radishes were introduced to other regions, such as ancient Egypt, Greece, and Rome, where they became popular not only as a food but also as a medicine. In ancient Greece and Rome, radishes were highly valued for their medicinal properties, believed to aid digestion, treat coughs and respiratory ailments, and even act as an aphrodisiac.

Can you eat it raw? Radishes are often eaten raw and can make a great addition to salads, sandwiches, or enjoyed as a healthy snack. They are known for their refreshing and spicy flavor and their crisp texture. When eaten raw, radishes can provide a variety of health benefits, such as being low in calories, high in fiber, and rich in vitamin C.

Growing and harvesting Radishes are easy to grow and can be planted both in the spring and fall. They prefer well draining soil and can tolerate both full sun and partial shade. When planting radishes, it is essential to keep the soil moist and avoid over watering, as this can lead to rotting. Radishes are usually ready to be harvested within 3-4 weeks of planting, depending on the variety. To harvest, gently pull the radish from the soil, being careful not to damage the roots. The ideal size for a radish is around 1 inch in diameter, with a firm and smooth texture.

Ripeness To determine if a radish is ripe, look for a firm and crisp texture with a bright color. The skin should be smooth, with no blemishes or soft spots, and the leaves should be green and vibrant.

Spoilage Radishes that have spoiled will have a soft, mushy texture and may develop brown or black spots. They may also have a sour or unpleasant smell.

Storing To store radishes, remove the leaves and place them in a plastic bag or airtight container in the refrigerator. Radishes can last up to a week when stored this way, but it is recommended to consume them as soon as possible for optimal freshness.

Preserving If you want to preserve radishes for longer periods, pickling is an excellent option. Simply soak the radishes in vinegar and saltwater to retain their flavor and texture. Pickled radishes can be used as a condiment or added to sandwiches and salads.

RADISH SALAD

Serves: 2
Prep Time: 10 minutes
Cook Time: 0

1 bunch of radishes, thinly sliced
15 ml extra virgin olive oil
5 ml fresh lemon juice
Salt and pepper, to taste

Directions

1. Wash and trim the radishes, then slice them thinly using a sharp knife or mandoline slicer. Place the sliced radishes in a large mixing bowl.
2. Drizzle the olive oil and lemon juice over the radishes, and season with salt and pepper.
3. Toss everything together until the radishes are evenly coated.
4. Serve the radish salad immediately.

Notes
Use a mandolin slicer for evenly and thinly sliced radishes.

ROASTED RADISH

Serves: 2
Prep Time: 10 minutes
Cook Time: 25 minutes

1 bunch of radishes, halved
30 ml extra virgin olive oil
Salt and pepper, to taste

Directions

1. Preheat the oven to 200°C (400°F). Line a baking sheet with parchment paper.
2. Wash and trim the radishes, then cut them in half lengthwise. Place the halved radishes in a large mixing bowl.
3. Drizzle the olive oil over the radishes, and season with salt and pepper. Toss everything together until the radishes are evenly coated.
4. Spread the radishes in a single layer on the prepared baking sheet.
5. Roast the radishes in the preheated oven for 20-25 minutes, or until they are tender and lightly browned.

Notes
Enhance the flavor by sprinkling the roasted radishes with chopped fresh herbs such as parsley, thyme, or chives before serving.

AIR FRIED RADISH CHIPS

Serves: 4
Prep Time: 10 minutes
Cook Time: 10 minutes

1 bunch of radishes, thinly sliced
30 ml extra virgin olive oil
Salt and pepper, to taste

Directions

1. Preheat the air fryer to 200°C (400°F).
2. Wash and trim the radishes, then slice them thinly using a mandolin slicer.
3. In a large mixing bowl, drizzle the olive oil over the radishes, and season with salt and pepper. Toss everything together until the radishes are evenly coated.
4. Spread the radishes on aluminum foil in a single layer in the air fryer basket.
5. Air fry the radishes in the preheated air fryer for 8-10 minutes, or until they are crispy and golden brown.

Notes
Add extra flavor by incorporating a pinch of garlic powder or smoked paprika into the seasoning mixture.

RUTABAGA

History Rutabaga, also known as swede, is believed to have originated in Scandinavia in the 17th century by crossing turnips with cabbage. The vegetable's name comes from the Swedish word "rotabagge," which translates to "root ram." Rutabaga quickly gained popularity in Northern Europe, where it became a staple food during long winters due to its hardiness and nutritional value.

Can you eat it raw? Although rutabaga can be eaten raw, it is most commonly cooked to improve its flavor and texture. Raw rutabaga has a slightly bitter taste and a tough, fibrous texture that can be challenging to digest. Cooking rutabaga not only improves its flavor but also softens its texture, making it more palatable.

Growing and harvesting Rutabaga is a root vegetable that grows best in cool weather and is typically harvested in the fall. It is usually planted in late spring or early summer and takes about three to four months to mature. Rutabaga requires well drained soil and regular watering to thrive. It is a hardy crop that can withstand frost and even tastes sweeter after a frost.

Ripeness A ripe rutabaga will have a smooth, unblemished skin with a deep purple top and creamy white bottom. It should feel heavy for its size and have a firm texture.

Spoilage Spoiled rutabaga will have a soft texture, brown spots, or wrinkles on the skin. It may also have a strong odor or mold. If you notice any of these signs, discard the rutabaga as it may be unsafe to consume.

Storing Rutabaga should be stored in a cool, dry place such as a pantry or root cellar. It can also be stored in the refrigerator for up to a week. Make sure to store rutabaga away from fruits and vegetables that give off ethylene gas, such as apples, bananas, and tomatoes, as this can cause it to spoil faster.

Peserving Rutabaga can be blanched and frozen for long term storage. To do this, peel and chop the rutabaga into small pieces, blanch in boiling water for 2-3 minutes, then immediately transfer to an ice bath to stop the cooking process. Once cooled, transfer to a freezer safe container and store in the freezer for up to 8 months.

ROASTED RUTABAGA

Serves: 2
Prep Time: 10 minutes
Cook Time: 25 minutes

1 large rutabaga, peeled and diced
15 ml extra virgin olive oil
Fresh lemon juice (optional)
Salt and pepper, to taste

Directions

1. Preheat the oven to 200°C (400°F).
2. In a large bowl, toss the rutabaga pieces with olive oil, lemon juice, salt, and pepper until well coated.
3. Spread the rutabaga pieces in a single layer on a baking sheet.
4. Roast in the preheated oven for 20-25 minutes, or until tender and lightly browned, stirring once or twice during cooking.
5. Serve hot

Notes

To enhance the flavor, consider adding minced garlic to the olive oil mixture.

RUTABAGA SOUP

Serves: 2
Prep Time: 10 minutes
Cook Time: 35 minutes

2 large rutabaga, peeled and diced
30 ml extra virgin olive oil
300 ml water or stock
Salt and pepper, to taste

Directions

1. In a large pot, heat the olive oil over medium heat.
2. Add the diced rutabaga and sauté for 5-7 minutes, or until lightly browned.
3. Add the water or stock and bring to a boil.
4. Reduce the heat and simmer for 20-25 minutes, or until the vegetables are tender.
5. Use an immersion blender or transfer the soup to a blender and blend until smooth.
6. Season with salt and pepper to taste.
7. Garnish with fresh herbs, croutons, or a drizzle of cream, if desired.
8. Serve hot.

Notes

For extra depth of flavor, try adding diced onion, diced carrots, and minced garlic to the sauté step.

AIR FRIED RUTABAGA FRIES

Serves: 2
Prep Time: 10 minutes
Cook Time: 20 minutes

1 large rutabaga, peeled and cut into fries
30 ml extra virgin olive oil
Salt and pepper, to taste

Directions

1. Preheat the air fryer to 200°C (400°F).
2. In a large bowl, toss the rutabaga fries with olive oil, salt, and pepper until well coated.
3. Spread the rutabaga fries on aluminum foil in a single layer in the air fryer basket.
4. Cook for 15-20 minutes, or until golden and crispy, shaking the basket occasionally during cooking.
5. Garnish with chopped fresh herbs, grated Parmesan cheese, or your favorite dipping sauce, if desired.
6. Serve hot

Notes

For added flavor, sprinkle the fries with a pinch of paprika or cayenne pepper before cooking.

SHALLOT

History Shallots are a member of the allium family, which also includes onions, garlic, and scallions. They are believed to have originated in Central or Southeast Asia and have been cultivated for thousands of years. Ancient Egyptians were known to use shallots in their cuisine, and they were also popular in ancient Rome.

Can you eat it raw? Yes, shallots can be eaten raw. They have a milder flavor than onions and are often used in salads, dressings, or as a garnish. Raw shallots can be sliced thinly or minced and used to add a subtle onion like flavor to dishes.

Growing and harvesting Shallots are usually planted in the fall for a spring harvest. The bulbs are planted about 1-2 inches deep and 4-6 inches apart in well draining soil. Shallots require full sun and regular watering but do not like to be waterlogged. The bulbs will form clusters of small bulbs, or "cloves," that can be harvested when the leaves turn yellow and begin to wither. Once harvested, shallots should be cured for a few days in a dry, well ventilated area before storage.

Ripeness Shallots are ripe when the leaves have turned yellow and begin to wither. The bulbs should be firm and have a papery skin. If the leaves have completely withered and fallen over, the shallots may be overripe and more prone to spoilage.

Spoilage Spoiled shallots will have a soft or mushy texture, a strong odor, or may have started to sprout. Mold or signs of rotting are also indicators of spoiled shallots. Once spoiled, shallots should be discarded as they can pose a risk of food-borne illness.

Storing Shallots should be stored in a cool, dry place with good air circulation. They can be stored at room temperature for up to a week or in the refrigerator for up to a month. To prevent moisture buildup, shallots should be stored in a paper bag or mesh bag.

Preserving Shallots can be preserved by pickling them in vinegar. Simply slice the shallots thinly and place them in a jar with vinegar, sugar, and spices. Seal the jar and let it sit in the refrigerator for at least 30 minutes to 24 hours before using. Pickled shallots can be used as a condiment or topping for sandwiches, salads, and more.

SHALLOT CONFIT

Serves: 6
Prep Time: 10 minutes
Cook Time: 90 minutes

450 g shallots, peeled and sliced
30 ml extra virgin olive oil
15 ml red wine vinegar (optional)
240 ml water or stock
Salt and pepper, to taste

Directions

1. Preheat the oven to 150°C (300°F).
2. In a large oven safe skillet, combine shallots, olive oil, red wine vinegar, water or stock, salt, and pepper.
3. Cover the skillet with foil and bake for 1 hour, stirring occasionally.
4. Remove the foil and bake for an additional 30 minutes, or until the shallots are soft and caramelized.
5. Serve as a side dish or as a topping for meat or vegetables.

Notes
Shallot confit can be stored in the refrigerator for up to a week in an airtight container.

SHALLOT VINAIGRETTE

Makes: 240 ml
Prep Time: 10 minutes
Cook Time: 0

120 ml finely chopped shallots
60 ml extra virgin olive oil
30 ml red wine vinegar
Salt and pepper, to taste

Directions

1. In a medium bowl, whisk together shallots, olive oil, vinegar, salt, and pepper until well combined.
2. Use immediately or store in an airtight container in the refrigerator for up to a week.

Notes
Shallot vinaigrette can be used as a salad dressing, marinade, or dipping sauce. For a smoother texture, blend the vinaigrette in a food processor or blender.

AIR FRIED SHALLOTS

Serves: 4
Prep Time: 5 minutes
Cook Time: 10 minutes

225 g shallots, peeled and sliced
30 ml extra virgin olive oil
Salt and pepper, to taste

Directions

1. Preheat the air fryer to 200°C (400°F).
2. In a large bowl, toss the shallots with olive oil, salt, and pepper until well coated.
3. Place the shallots in the air fryer basket in a single layer.
4. Air fry for 8-10 minutes, shaking the basket halfway through cooking, until the shallots are golden brown and crispy.
5. Repeat with remaining shallots if necessary.
6. Serve as a snack or as a topping for burgers, sandwiches, or salads.

Notes
Air fried shallots can also be seasoned with spices such as garlic powder, paprika, or cayenne pepper.

SPINACH

History Spinach has a long and interesting history dating back to ancient Persia (modern-day Iran). It is believed to have been cultivated over 2,000 years ago and was introduced to other parts of the world, including India and China. The vegetable made its way to Europe in the 11th century through trade routes, and by the 12th century, it had become a popular leafy green in the Mediterranean region. Spinach was also popular among sailors due to its high vitamin C content, which prevented scurvy during long voyages.

Can you eat it raw? Spinach can be eaten raw or cooked. Raw spinach is a popular addition to salads and smoothies.

Growing and harvesting Spinach is a cool weather crop that thrives in temperatures between 15-20°C (60-68°F). It can be grown from seeds or seedlings and prefers well drained, fertile soil with a neutral pH level. Spinach plants typically take between 30-45 days to mature and are harvested when they are 4-6 inches tall and have 4-6 leaves. The leaves are harvested by cutting them off at the base of the stem, and new leaves will grow back from the crown.

Ripeness Fresh spinach leaves should be bright green, with no yellow or brown spots. The leaves should be firm to the touch and not wilted. Yellow or brown spots indicate that the leaves are past their prime and may not taste as good.

Spoilage Spinach that has spoiled will have wilted leaves that are yellow or brown in color. The leaves will have a slimy texture and a foul odor. If you notice any of these signs, discard the spinach as it may be harmful to consume.

Storing Fresh spinach should be stored in a plastic bag or container in the refrigerator. It is best to use spinach within 3-5 days of purchase. Do not wash spinach before storing as the excess moisture can cause it to spoil more quickly. If the spinach starts to wilt, you can try reviving it by placing it in a bowl of ice water for a few minutes before using.

Preserving The simplest way to preserve spinach is to blanch it before freezing. To blanch, bring a pot of water to a boil and submerge the spinach leaves for 1-2 minutes. Remove the leaves from the boiling water and place them in ice water to stop the cooking process. Once cooled, drain the leaves and place them in a freezer bag or container. Frozen spinach can be stored for up to 8 months.

SAUTÉED SPINACH

Serves: 2
Prep Time: 10 minutes
Cook Time: 5 minutes

225 g spinach, washed and dried
15 ml extra virgin olive oil
Fresh lemon juice (optional)
Salt and pepper, to taste

Directions

1. Remove any tough stems from the spinach leaves before cooking.
2. Heat the olive oil in a large skillet over medium heat.
3. Add the spinach leaves and stir until wilted, about 2-3 minutes.
4. Drizzle with lemon juice and season with salt and pepper, to taste.
5. Transfer to a serving platter and serve hot.

Notes
For added flavor, sprinkle some grated Parmesan cheese over the sautéed spinach before serving.

SPINACH SALAD

Serves: 1
Prep Time: 10 minutes
Cook Time: 0 minutes

60 g spinach, washed and dried
10 ml extra virgin olive oil
15 ml balsamic vinegar
Salt and pepper, to taste

Directions

1. Arrange the spinach leaves on a large serving platter.
2. In a small bowl, whisk together the olive oil and balsamic vinegar to make the dressing.
3. Drizzle the dressing over the spinach leaves and season with salt and pepper, to taste.
4. Serve immediately, spinach starts to wilt once in contact with the acid.

Notes
For added texture and flavor, top the spinach salad with sliced strawberries, crumbled feta cheese, and toasted nuts such as almonds or walnuts.

AIR FRIED SPINACH CRISPS

Serves: 1
Prep Time: 10 minutes
Cook Time: 7 minutes

125 g spinach, washed and dried
30 ml extra virgin olive oil
Salt and pepper, to taste

Directions

1. Pat the spinach leaves dry with a paper towel before tossing with olive oil.
2. Preheat the air fryer to 190°C (375°F).
3. In a large bowl, toss the spinach leaves with olive oil and season with salt and pepper.
4. Arrange the spinach leaves on aluminum foil in a single layer in the air fryer basket and place stainless steel utensils on top of spinach to prevent movement while cooking.
5. Cook for 5-7 minutes, flipping the leaves halfway through, until they are crispy and golden brown.
6. Transfer the spinach chips to a serving platter and serve immediately.

Notes
Experiment with different seasonings such as garlic powder, cumin, or chili flakes to add some spice to the spinach chips.

SQUASH

History Squash has a rich history dating back thousands of years in the Americas. The oldest evidence of squash cultivation comes from Oaxaca, Mexico, where remains of domesticated squash dating back to 8000 BC have been found. Native American tribes across North and South America cultivated squash as a staple food source, using various types for their flesh, seeds, and fibers. Squash became an important crop for European settlers and was later introduced to other parts of the world, including Africa and Asia.

Can you eat it raw? While squash is typically cooked before eating, some varieties can be consumed raw, such as summer squash and zucchini. These types of squash have a mild, slightly sweet flavor and a tender texture that makes them great for adding to salads or for dipping in hummus or other dips.

Growing and harvesting Squash is a warm weather crop that requires plenty of sunshine and well drained soil. It can be planted in the spring after the danger of frost has passed and harvested in the late summer or early fall when the fruits are mature. Squash plants can produce a large yield, so it is important to harvest the fruits frequently to prevent them from becoming overripe.

Ripeness The ripeness of squash can be determined by its size, color, and texture. Most varieties of squash are ready to harvest when they are full sized and have a deep, rich color. The skin should be firm and free from blemishes, and the stem should be dry and brown. When pressed gently, the skin should yield slightly but should not be soft or mushy.

Spoilage Squash that has spoiled will typically have a soft, mushy texture and a foul smell. The skin may also appear wrinkled or discolored, and there may be visible signs of mold or fungus.

Storing Squash should be stored in a cool, dry place away from direct sunlight. Do not store squash in the refrigerator, as this can cause it to spoil more quickly. If stored properly, most varieties of squash can last for several weeks.

Preserving Squash can be easily preserved by freezing. To freeze squash, wash and slice it into 1/2 inch rounds or cubes. Blanch the squash in boiling water for 3 minutes, then immediately transfer to an ice bath to stop the cooking process. Drain the squash and pack it into airtight containers or freezer bags. Squash can be frozen for up to 8 months.

SAUTEED SQUASH

Serves: 2
Prep Time: 5 minutes
Cook Time: 15 minutes

1 medium acorn squash, sliced into rounds
30 ml extra virgin olive oil
Fresh lemon juice (optional)
Salt and pepper, to taste

Directions

1. In a large skillet, heat the olive oil over medium heat.
2. Add the sliced squash and cook for 10-15 minutes or until tender and slightly browned.
3. Squeeze lemon juice over the squash and season with salt and pepper, to taste.

Notes

For added flavor, garnish with freshly grated Parmesan cheese or add a pinch of red pepper flakes for a little kick.

ROASTED SQUASH

Serves: 2
Prep Time: 5 minutes
Cook Time: 25 minutes

1 medium squash, sliced into rounds
30 ml extra virgin olive oil
Balsamic glaze (optional)
Salt and pepper, to taste

Directions

1. Preheat the oven to 200°C (400°F).
2. In a large bowl, toss the sliced squash with olive oil and balsamic.
3. Spread the squash out in a single layer on a baking sheet.
4. Season with salt and pepper.
5. Roast in the oven for 20-25 minutes or until the squash is tender and caramelized.

Notes

Experiment with different herbs like rosemary or thyme for added flavor. You can also try adding a sprinkle of feta cheese before serving.

AIR FRIED SQUASH FRIES

Serves: 4
Prep Time: 5 minutes
Cook Time: 12 minutes

1 medium squash, sliced into thin fries
30 ml extra virgin olive oil
Salt and pepper, to taste

Directions

1. Preheat the air fryer to 200°C (400°F).
2. In a large bowl, toss the squash fries with olive oil, salt, and pepper.
3. Place the squash fries on aluminum foil in the air fryer basket, in a single layer.
4. Cook for 10-12 minutes or until the squash fries are crispy and golden brown.

Notes

Serve with a dipping sauce like aioli, ranch dressing, or marinara sauce for an extra burst of flavor.

SUNCHOKE

History Sunchokes, also known as Jerusalem artichokes, are native to North America and were originally cultivated by Native Americans. They were an important food source for many tribes long before the arrival of Europeans in the New World. Despite the name "Jerusalem," the origin of the name is not clear.

Can you eat it raw? Yes, sunchokes can be eaten raw. They have a crunchy texture and a slightly sweet, nutty flavor that makes them a great addition to salads or vegetable platters. However, some people may experience digestive issues when eating them raw, so it's best to start with a small amount and see how your body reacts.

Growing and harvesting Sunchokes are grown from tubers, which are planted in the spring after the last frost. They prefer a sunny location with well drained soil, and can be grown in containers or directly in the ground. They require regular watering and weeding, and the plants can grow up to 10 feet tall with yellow sunflower like flowers. Sunchokes are harvested in the fall after the foliage has died back, and the tubers are dug up and stored.

Ripeness Sunchokes are typically harvested in the fall when the foliage has died back, and the tubers are mature and ready to eat. They should be firm to the touch and not soft or mushy.

Spoilage Sunchokes can spoil if they are not stored properly. Signs of spoilage include mold, soft spots, or a bad odor. If the tubers have started to sprout, they are still safe to eat but may have a slightly bitter taste.

Storing Sunchokes should be stored in a cool, dry place, such as a root cellar or refrigerator. They can be stored in a paper or mesh bag to allow for air circulation, and should be used within a few weeks of harvest.

Preserving Sunchokes can be preserved by freezing. Simply blanch the tubers for 3-4 minutes, plunge them into ice water to stop the cooking process, and then freeze in an airtight container for up to 6 months.

ROASTED SUNCHOKES

Serves: 1
Prep Time: 10 minutes
Cook Time: 25 minutes

125 g sunchokes, washed, peeled, and small diced
15 ml extra virgin olive oil
Fresh lemon juice (optional)
Salt and pepper, to taste

Directions

1. Preheat the oven to 200°C (400°F).
2. In a large bowl, toss the sunchokes with the olive oil, lemon juice, salt, and pepper until evenly coated.
3. Spread the sunchokes out on a baking sheet in a single layer.
4. Roast for 20-25 minutes, until tender and golden brown, stirring occasionally.

Notes

Consider trying different herbs like rosemary or thyme for different flavor profiles

SAUTÉED SUNCHOKES

Serves: 1
Prep Time: 10 minutes
Cook Time: 12 minutes

125 g sunchokes, washed, peeled, and sliced into rounds
5 ml extra virgin olive oil
Balsamic glaze (optional)
Salt and pepper, to taste

Directions

1. Heat the olive oil in a large skillet over medium-high heat.
2. Add the sunchokes and cook for 8-10 minutes, stirring occasionally, until browned and tender.
3. Add the balsamic and stir to coat the sunchokes.
4. Cook for an additional 1-2 minutes, until the vinegar has reduced and the sunchokes are glazed.
5. Season with salt and pepper, to taste.

Notes

For a sweeter version, use honey or maple syrup instead of balsamic glaze.

AIR FRIED SUNCHOKES

Serves: 1
Prep Time: 10 minutes
Cook Time: 15 minutes

125 g sunchokes, washed, peeled, and small diced
30 ml extra virgin olive oil
Salt and pepper, to taste

Directions

1. Preheat the air fryer to 200°C (400°F).
2. In a large bowl, toss the sunchokes with the olive oil, salt, and pepper until evenly coated.
3. Place the sunchokes on aluminum foil in the air fryer basket in a single layer.
4. Air fry for 12-15 minutes, until crispy and golden brown, shaking the basket halfway through.

Notes

Enhance the flavor with a squeeze of lemon juice or a sprinkle of parmesan cheese. Try adding minced garlic and fresh rosemary leaves to the olive oil before cooking

SWEET POTATO

History Sweet potatoes have a long history that dates back over 5,000 years in South America. It is believed that the first sweet potatoes were domesticated in the Andes Mountains of Peru. From there, they were brought to other parts of South America, the Caribbean, and Central America. In the 16th century, Spanish explorers brought sweet potatoes to Europe, where they became popular among the wealthy as a delicacy. Sweet potatoes were later introduced to North America by European settlers and were grown by Native American tribes.

Can you eat it raw? While sweet potatoes can technically be eaten raw, they are not typically consumed this way due to their tough and starchy texture. Raw sweet potatoes can also be difficult to digest and may cause gastrointestinal discomfort. It is recommended to cook sweet potatoes before consuming.

Growing and harvesting Sweet potatoes are grown in warm climates and require a long growing season of around 100-150 days. They are usually planted in the spring and harvested in the fall once the leaves begin to yellow and die back. The sweet potatoes are carefully dug up from the ground, taking care not to damage the skin, and then cured for a few weeks in a warm and humid environment. Curing improves their flavor and texture and helps to heal any cuts or bruises on the skin.

Ripeness When selecting sweet potatoes, look for ones that feel firm and have smooth, unblemished skin. The color can vary depending on the variety, but ripe sweet potatoes should have a uniform color and no signs of sprouting.

Spoilage Spoiled sweet potatoes will have soft spots, mold, or a foul smell. Do not consume sweet potatoes that exhibit any of these signs of spoilage.

Storing Sweet potatoes should be stored in a cool, dry place, away from direct sunlight. Do not store sweet potatoes in the refrigerator, as the cold temperature can cause the starch to convert to sugar, altering their taste and texture. When stored properly, sweet potatoes can last for several weeks to a few months.

Preserving Sweet potatoes can be preserved by freezing. To freeze, cook the sweet potatoes, let them cool, and then freeze in an airtight container.

ROASTED SWEET POTATOES

Serves: 1
Prep Time: 10 minutes
Cook Time: 25 minutes

1 medium sweet potatoes, peeled and cubed
15 ml extra virgin olive oil
Fresh lemon juice (optional)
Salt and pepper, to taste

Directions

1. Preheat the oven to 220°C (430°F). Line a baking sheet with parchment paper.
2. In a medium bowl, toss the potatoes with the oil until evenly coated.
3. Spread the sweet potatoes in a single layer.
4. Roast the sweet potatoes for 20-25 minutes, or until tender and golden brown, stirring halfway through.
5. In a small bowl, whisk together the lemon juice, salt, and pepper.
6. Drizzle the lemon mixture over the roasted sweet potatoes and serve.

Notes

For a more robust flavor, consider adding a minced garlic or thyme to the lemon mixture.

MASHED SWEET POTATO

Serves: 1
Prep Time: 10 minutes
Cook Time: 20 minutes

1 medium sweet potato, peeled and cut into small pieces
15 ml butter or extra virgin olive oil
Apple cider vinegar (optional)
Salt and white pepper, to taste

Directions

1. In a large pot, bring the sweet potatoes and enough water to cover to a boil. Reduce the heat and simmer for 15-20 minutes, or until tender.
2. Drain the sweet potatoes and return them to the pot.
3. Add the butter or oil, apple cider vinegar, salt, and pepper. Mash the sweet potatoes with a potato masher or immersion blender until smooth and creamy.
4. Adjust the seasoning as needed and serve hot.

Notes

For a creamier texture, add a splash of milk or cream while mashing.

AIR FRIED SWEET POTATO FRIES

Serves: 2
Prep Time: 10 minutes
Cook Time: 15 minutes

1 medium sweet potato, peeled and cut into thin fries
30 ml extra virgin olive oil
Salt and pepper, to taste
White wine vinegar (optional)

Directions

1. Preheat the air fryer to 220°C (430°F).
2. In a medium bowl, toss the sweet potato fries with the olive oil until evenly coated. Season with salt and pepper.
3. Arrange the sweet potato fries on aluminum foil in a single layer in the air fryer basket.
4. Cook for 12-15 minutes, or until crispy and golden brown, shaking the basket halfway through.
5. Serve hot with your favorite dipping sauce.

Notes

For an extra burst of flavor, sprinkle the fries with your favorite seasoning blend before cooking

SWISS CHARD

History Swiss chard, also known as silverbeet or spinach beet, is a leafy green vegetable that belongs to the Chenopodioideae family. It is closely related to beets and spinach and is believed to have originated in the Mediterranean region, where it has been cultivated for thousands of years. Swiss chard was introduced to North America in the late 19th century and has since become a popular vegetable worldwide.

Can you eat it raw? Yes, Swiss chard can be eaten raw, but it is often cooked before consumption. When eaten raw, it is best to use young, tender leaves in salads or as a garnish. The mature leaves can have a tough texture and a slightly bitter taste.

Growing and harvesting Swiss chard is a cool season crop that can be grown in spring or fall. It prefers well draining soil that is rich in organic matter. The seeds should be planted 1/2 inch deep and 1 inch apart. The plants should be thinned to 6-12 inches apart once they are about 3 inches tall. Swiss chard is typically ready for harvest 55-60 days after planting. The outer leaves can be harvested by cutting them off at the base of the stem, or the entire plant can be harvested by cutting it at ground level.

Ripeness The ripeness of Swiss chard can be determined by the size of the leaves. Young leaves are smaller and more tender, while older leaves are larger and tougher. Look for leaves that are bright green and crisp.

Spoilage Spoiled Swiss chard will have wilted leaves, yellow or brown spots, or a slimy texture. It may also have a sour smell. It is best to discard any Swiss chard that has spoiled.

Storing Swiss chard should be stored in the refrigerator. To keep it fresh, wrap the leaves in a damp paper towel and place them in a plastic bag. It can be stored in the refrigerator for up to 5 days.

Preserving The simplest way to preserve Swiss chard is by blanching and freezing it for later use. To blanch Swiss chard, bring a pot of water to a boil and add the leaves. Cook for 2-3 minutes, then immediately place them in ice water to stop the cooking process. Once the leaves have cooled, squeeze out any excess water and place them in a freezer safe bag. They can be stored in the freezer for up to 8 months.

SAUTÉED SWISS CHARD

Serves: 4
Prep Time: 10 minutes
Cook Time: 10 minutes

1 bunch Swiss chard (about 225g)
30 ml extra virgin olive oil
Fresh lemon juice (optional)
Salt and pepper, to taste

Directions

1. Wash the Swiss chard thoroughly and remove the tough stems. Chop the leaves into bite sized pieces.
2. Heat the olive oil in a large skillet over medium heat.
3. Add the Swiss chard to the skillet and sauté for 3-5 minutes, stirring occasionally, until the leaves are wilted and tender.
4. Drizzle the lemon juice over the Swiss chard and season with salt and pepper.
5. Transfer the Swiss chard to a serving platter and serve hot.

Notes

Add red pepper flakes or a bit of grated Parmesan cheese on top for an extra burst of flavor.

STEAMED SWISS CHARD

Serves: 4
Prep Time: 10 minutes
Cook Time: 5 minutes

1 bunch Swiss chard (about 225g)
15 ml extra virgin olive oil
Apple cider vinegar (optional)
Salt and pepper, to taste

Directions

1. Wash the Swiss chard thoroughly and remove the tough stems. Chop the leaves into bite sized pieces.
2. Fill a pot with 25-50 ml of water and bring to a boil. Place a steamer basket in the pot and add the Swiss chard to the basket.
3. Cover the pot and steam the Swiss chard for 3-5 minutes, until the leaves are wilted and tender.
4. Remove the Swiss chard from the steamer basket and transfer it to a serving platter.
5. Drizzle the olive oil and white wine vinegar over the Swiss chard and season with salt and pepper.
6. Serve hot.

Notes

Serve with a lemon wedge for a fresh, zesty flavor.

AIR FRIED SWISS CHARD CHIPS

Serves: 4
Prep Time: 10 minutes
Cook Time: 6 minutes

1 bunch Swiss chard (about 225g)
60 ml extra virgin olive oil
Salt and pepper, to taste

Directions

1. Wash the Swiss chard thoroughly and remove the tough stems. Tear the leaves into desird size pieces.
2. Preheat the air fryer to 200°C (400°F).
3. In a large bowl, toss the Swiss chard leaves with olive oil, salt, pepper until evenly coated.Working in batches, place the Swiss chard leaves in a single layer in the air fryer basket and cover in stainless steel utensils to prevent movement while cooking.
4. Air fry for 4-6 minutes, or until the leaves are crispy and lightly browned.
5. Remove the Swiss chard chips from the air fryer and garnish with additional salt and pepper, if desired.

Notes

Experiment with different seasonings, such as garlic powder, onion powder, or a cayenne pepper.

TOMATO

History The tomato is native to western South America and was first cultivated by the Incas as early as 700 AD. The tomato plant was later domesticated by the Aztecs and spread throughout Central America and Mexico. Spanish explorers brought the tomato to Europe in the 16th century, and it was initially grown as an ornamental plant due to its bright colors. The tomato eventually became a popular ingredient in Mediterranean cuisine, with the first recorded use of tomatoes in Italian cuisine dating back to the late 17th century.

Can you eat it raw? Yes, the tomato can be eaten raw and is commonly used in salads and sandwiches.

Growing and harvesting Tomatoes are typically grown in warm, sunny climates and are often cultivated in greenhouses. They can be grown in soil or hydroponically (without soil). The plants are usually started from seeds and are transplanted to the field or greenhouse once they reach a certain size. Tomatoes are harvested when they are ripe, which is determined by their color and firmness.

Ripeness A ripe tomato should be firm but slightly soft to the touch. The color of the tomato should be uniformly red (or yellow, orange, or pink, depending on the variety). If the tomato is still green, it is not yet ripe. Overripe tomatoes will be soft and mushy, and the skin may be wrinkled.

Spoilage A spoiled tomato will have a soft, mushy texture and may have mold or other signs of decay. It may also have a sour or unpleasant odor. It is best to discard spoiled tomatoes as they may cause foodborne illness if consumed.

Storing Tomatoes should be stored at room temperature, out of direct sunlight, and with the stem side down. They can be stored in a single layer or in a shallow container. Do not store tomatoes in the refrigerator, as this can cause them to lose flavor and become mealy. To extend their shelf life, store ripe tomatoes in a cool, dry place with good air circulation.

Preserving One simple way to preserve tomatoes is to freeze them. To do this, blanch the tomatoes in boiling water for a minute or two, then plunge them into ice water to stop the cooking process. Remove the skins and any cores or seeds, then freeze the tomatoes in a single layer on a baking sheet. Once frozen, transfer them to a freezer safe container.

ROASTED TOMATO SAUCE

Serves: 2
Prep Time: 10 minutes
Cook Time: 1-3 hours

4 large San Marzano tomatoes, halved
30 ml extra virgin olive oil
Balsamic reduction (optional)
240 ml water or stock
Salt and pepper, to taste

Directions

1. Preheat the oven to 200°C (400°F).
2. Arrange the tomatoes on a baking sheet. Drizzle with olive oil and balsamic.
3. Roast in the oven for 20-30 minutes, until the tomatoes are soft and slightly charred.
4. Transfer the roasted tomatoes to a blender or food processor. Add liquid and blend until smooth.
5. Transfer the sauce to a pot and heat over medium heat until hot. Season with salt and pepper, to taste.

Notes

For a creamier sauce, add heavy cream or coconut cream before serving. Serve with pasta, as a pizza base, or roasted vegetables. Try roasting garlic along with the tomatoes.

HEIRLOOM TOMATO SALAD

Serves: 1
Prep Time: 10 minutes
Cook Time: 0 minutes

2 medium heirloom tomatoes, sliced
30 g fresh basil leaves, chopped
15 ml extra virgin olive oil
15 ml balsamic glaze or fresh lemon juice
Salt and pepper, to taste

Directions

1. Arrange the sliced tomatoes on a serving plate.
2. Sprinkle the chopped basil over the tomatoes.
3. Drizzle with olive oil and balsamic or juice.
4. Season with salt and pepper, to taste.

Notes

For added flavor, top the salad with fresh mozzarella or crumbled feta cheese and sliced avocado. This salad is best served immediately after preparation.

AIR FRIED TOMATO SLICES

Serves: 1
Prep Time: 5 minutes
Cook Time: 8 minutes

2 medium campari tomatoes, sliced
15 ml extra virgin olive oil
Salt and pepper, to taste

Directions

1. Preheat the air fryer to 200°C (400°F).
2. Arrange the tomato slices on aluminum foil in a single layer in the air fryer basket.
3. Drizzle with olive oil and season with salt and pepper.
4. Air fry for 6-8 minutes, until the tomatoes are slightly charred and tender.
5. Serve as a side dish or use as a topping for sandwiches, burgers, or pizzas.

Notes

For added flavor, sprinkle the tomatoes with grated Parmesan cheese or chopped herbs before air frying. The air fryer time may vary depending on the model and thickness of the tomato.

TURNIP

History Turnips have a long history of cultivation and have been a popular food crop for over 4,000 years. Originating in Europe and Asia, turnips were widely consumed by the Greeks and Romans. The ancient Greeks believed that turnips had mystical powers, and the Romans valued them for their medicinal properties. Turnips were introduced to the United States in the 1600s by European settlers and have since become a staple ingredient in many cuisines around the world.

Can you eat it raw? While turnips are more commonly cooked, they can be eaten raw. Raw turnips have a slightly bitter and peppery taste and can be grated or sliced thinly to add a refreshing crunch to salads or slaws.

Growing and harvesting Turnips are cool weather crops that require full sun exposure and well draining soil to grow. They are usually planted in early spring or fall and take 2-3 months to mature. Turnips prefer temperatures between 50-60°F for optimal growth. When harvesting, wait until the roots reach 2-3 inches in diameter, and the leaves are still tender and green. The greens can also be harvested earlier, but be sure to leave enough leaves to support the growth of the root.

Ripeness A ripe turnip has a firm and smooth root with a vibrant purple and white coloring. The leaves should be tender to the touch and green. If the root is soft or has brown spots, it may be overripe or spoiled.

Spoilage Spoiled turnips have a soft and mushy texture with a sour or unpleasant smell. Mold or brown spots on the skin are also signs of spoilage. If you notice any of these signs, discard the turnip.

Storing For optimal freshness, store turnips in a plastic bag or airtight container in the refrigerator for up to two weeks. Remove the leaves and store them separately. If you have harvested your own turnips, leave a bit of stem on the root to keep it fresher longer.

Preserving To freeze turnips, peel and slice them before blanching them in boiling water for 3-5 minutes. Drain and allow them to cool before placing them in freezer bags. Turnips can be kept frozen for up to eight months.

ROASTED TURNIPS

Serves: 2
Prep Time: 10 minutes
Cook Time: 30 minutes

2 medium turnips, peeled and cut into wedges
30 ml extra virgin olive oil
Balsamic glaze (optional)
Salt and pepper, to taste

Directions

1. Preheat the oven to 200°C (400°F).
2. In a bowl, toss the turnip wedges with the olive oil, balsamic, salt, and pepper until the turnips are evenly coated.
3. Arrange the turnips in a single layer on a baking sheet.
4. Roast the turnips in the preheated oven for 25-30 minutes, until golden brown and tender.

Notes

For added flavor, consider adding a sprinkle of grated Parmesan cheese on top before serving.

TURNIP SALAD

Serves: 1
Prep Time: 10 minutes
Cook Time: 0

2 medium turnips, peeled and thinly sliced
15 ml extra virgin olive oil
15 ml fresh lemon juice
Salt and pepper, to taste

Directions

1. In a large bowl, combine the sliced turnips and apple.
2. Drizzle the olive oil and lemon juice over the top of the salad.
3. Sprinkle the salt and pepper over the salad and toss until the ingredients are evenly coated.
4. Let the salad sit for 10 minutes before serving to allow the flavors to meld together.

Notes

Add crumbled feta or goat cheese for a creamy element, thinly sliced apples for freshness, or toasted almonds for crunch.

AIR FRIED TURNIP FRIES

Serves: 2
Prep Time: 10 minutes
Cook Time: 15 minutes

1 medium turnips, peeled and cut into thin strips
30 ml extra virgin olive oil
Salt and pepper, to taste

Directions

1. Preheat the air fryer to 200°C (400°F).
2. In a bowl, toss the turnip strips with the olive oil, salt, and pepper until the turnips are evenly coated.
3. Arrange the turnip strips on aluminum foil in a single layer in the air fryer basket.
4. Air fry the turnips for 12-15 minutes, shaking the basket halfway through the cooking time, until the turnips are crispy and golden brown.

Notes

Experiment with different seasonings, such as garlic powder, paprika, or cayenne pepper, to create a unique flavor profile.

WATERCRESS

History Watercress has a long and rich history of use in culinary and medicinal applications. It has been cultivated for over 3,000 years and was considered a superfood by the ancient Greeks, Romans, and Persians. It was also used in traditional Chinese medicine to treat respiratory and digestive disorders. Watercress was introduced to the United States by European settlers in the 1800s, and it has since been widely cultivated in the southeastern states.

Can you eat it raw? Watercress is a versatile vegetable that can be enjoyed raw or cooked. Its crisp texture and peppery taste make it an excellent addition to salads, sandwiches, and soups. However, when consuming raw watercress, it is crucial to wash the leaves thoroughly and trim the stems before consumption to remove any dirt or debris.

Growing and harvesting Watercress is an aquatic plant that grows best in shallow, fast-moving water, such as streams or springs. It can be grown in hydroponic systems or in soil, but it requires a constant supply of water to thrive. Watercress is typically harvested by hand, with workers standing in the water and cutting the stems at the base of the plant.

Ripeness When purchasing watercress, it is best to look for bright green leaves and firm stems. Avoid any wilted or yellowed leaves, as they indicate that the watercress is past its prime and may have a bitter taste.

Spoilage Watercress has a short shelf life and can quickly spoil if not stored properly. Signs of spoilage include wilting, yellowing, or a slimy texture. To prolong its freshness, it should be kept in the refrigerator in a plastic bag with a damp paper towel. Watercress should not be stored near fruits, such as apples or bananas, as they release ethylene gas, which can cause the watercress to spoil more quickly.

Storing To store watercress, it should be placed in a plastic bag with a damp paper towel to keep it fresh. Properly stored watercress can last up to four days in the refrigerator.

Preserving Watercress can be preserved by blanching it in boiling water for 30 seconds, then quickly cooling it in ice water. Once blanched, it can be frozen for up to six months, ensuring its availability all year round.

WATERCRESS SALAD

Serves: 1
Prep Time: 10 minutes
Cook Time: 0

1 bunch watercress (about 100g), washed and trimmed
15 ml extra virgin olive oil
15 ml balsamic glaze
Salt and pepper, to taste

Directions

1. In a large bowl, combine the washed and trimmed watercress, olive oil, balsamic, salt, and pepper.
2. Toss the watercress until it is well coated in the dressing.
3. Serve immediately.

Notes

To enhance the flavor, you can add some freshly cracked black pepper or a sprinkle of grated Parmesan cheese.

WATERCRESS SOUP

Serves: 1
Prep Time: 10 minutes
Cook Time: 20 minutes

1 bunch watercress (about 100g), washed and trimmed
30 ml extra virgin olive oil
200 ml water or stock
Salt and pepper, to taste

Directions

1. Heat the olive oil in a large pot over medium heat.
2. Add the washed and trimmed watercress, water, salt, and pepper to the pot and bring to a boil.
3. Reduce the heat and simmer for 15 minutes.
4. Use an immersion blender to puree the soup until smooth.
5. Serve the watercress soup hot.

Notes

You can garnish the soup with a dollop of plain Greek yogurt, a drizzle of olive oil, or some chopped fresh herbs like parsley or chives.

AIR FRIED WATERCRESS CHIPS

Serves: 1
Prep Time: 5 minutes
Cook Time: 7 minutes

1 bunch watercress (about 100g), washed and dried
15 ml extra virgin olive oil
Salt and pepper, to taste

Directions

1. Preheat the air fryer to 190°C (375°F).
2. In a large bowl, toss the washed and dried watercress with olive oil, salt, and peppr.
3. Spread the watercress on aluminum foil in a single layer in the air fryer basket and secure with stainless steel utensils.
4. Cook for 5-7 minutes, shaking the basket halfway through, until the watercress is crispy and lightly browned.
5. Serve the air fryer watercress chips immediately.

Notes

You can add some garlic powder, onion powder, or chili flakes to the watercress before cooking for some added flavor.

ZUCCHINI

History Zucchini, also known as courgette, is a summer squash that belongs to the Cucurbitaceae family. Its origins can be traced back to Central and South America, where it was domesticated by indigenous people over 7,000 years ago. Zucchini was introduced to Europe by Christopher Columbus during his explorations of the New World. It was then cultivated in Italy in the 19th century, where it gained popularity and became a staple ingredient in Italian cuisine.

Can you eat it raw? Yes, zucchinis can be eaten raw. They are a great addition to salads when sliced thinly or cut into sticks and served with a dip.

Growing and harvesting Zucchini plants thrive in warm weather and require well draining soil. The seeds are usually planted directly in the ground after the last frost date, in hills or rows. The plants grow quickly and need plenty of water to produce good yields. The first harvest can be expected within 45-50 days of planting, and the zucchinis should be harvested when they are about 6-8 inches long and 2 inches in diameter. Regular harvesting promotes continued production throughout the growing season.

Ripeness Ripe zucchinis should feel firm to the touch and have a shiny, smooth skin. They should not have any soft spots or blemishes. Overripe zucchinis may become tough and seedy, and are best used for cooking.

Spoilage Spoiled zucchinis will have soft spots, mold, or a bad smell. If any of these signs are present, they should not be consumed.

Storing Zucchinis should be stored in a cool, dry place, such as the refrigerator's crisper drawer. They should be wrapped in a paper towel or a cloth to absorb any moisture. Whole zucchinis can last up to one week in the refrigerator. Cut zucchini should be used within two to three days.

Preserving Zucchinis can be easily frozen for later use. Simply slice them and blanch them in boiling water for 2-3 minutes, then cool them down in an ice bath. Once cooled, place them in a freezer container and store them in the freezer for up to 8 months.

ROASTED ZUCCHINI

Serves: 2
Prep Time: 10 minutes
Cook Time: 20 minutes

1 medium zucchinis, sliced into rounds
5 ml extra virgin olive oil
5 ml balsamic glaze (optional)
Salt and pepper, to taste

Directions

1. Preheat the oven to 200°C (400°F).
2. In a large mixing bowl, combine the sliced zucchinis, olive oil, balsamic, salt, and pepper. Toss until well coated.
3. Arrange the zucchini rounds in a single layer on a baking sheet.
4. Roast in the preheated oven for 15-20 minutes, or until golden brown and tender.
5. Remove from the oven and serve immediately.

Notes
For added flavor, sprinkle zucchini with grated Parmesan cheese or fresh herbs before roasting.

PAN CHARRED ZUCCHINI RIBBONS

Serves: 1
Prep Time: 15 minutes
Cook Time: 10 minutes

1 zucchini, sliced longways on a mandolin
Salt and pepper, to taste
5 ml extra virgin olive oil

Directions

1. Heat a small amount of vegetable oil in a large nonstick skillet over medium-high heat.
2. Season both sides of the zucchini slices
3. Place the zucchini slices in the skillet and cook until slightly charred on both sides, about 3-4 minutes per side.
4. Continue to add a small amount of oil to the pan before each batch of zucchini slices.

Notes
Serve with a dollop of yogurt or sour cream and a sprinkle of fresh herbs for a refreshing twist.

AIR FRIED ZUCCHINI CHIPS

Serves: 2
Prep Time: 10 minutes
Cook Time: 12 minutes

1 medium zucchinis, sliced into rounds
15 ml extra virgin olive oil
Salt and pepper, to taste

Directions

1. Preheat the air fryer to 200°C (400°F).
2. In a large mixing bowl, combine the sliced zucchinis, olive oil, salt, and pepper. Toss until evenly coated.
3. Arrange the zucchini rounds on aluminum foil in a single layer in the air fryer basket.
4. Air fry for 10-12 minutes, or until golden brown and crispy.
5. Remove from the air fryer and serve immediately.

Notes
Experiment with different seasonings, such as garlic powder, onion powder, or paprika, for a flavorful twist on this simple snack. Try dipping into seasoned breadcrumbs before cooking.

SIMPLE AND SATISFYING GAZPACHO

While Gazpacho itself may not be a single vegetable, it is undoubtedly my favorite harmony of vegetables, and thus, I am elated to include this delightful dish in the book.

During my unforgettable journey to Guadalajara, Spain, I had the pleasure of meeting the Del Castillo family, a warm and loving group who instantly welcomed me into their home. As the sun beat down on the vibrant Spanish streets, the Del Castillo matriarch prepared a refreshing Gazpacho that captured the essence of Spanish summer in a single, delightful spoonful. I marveled at the harmony of flavors, the perfect blend of tomatoes, peppers, and cucumbers, and couldn't help but feel a sense of belonging in that moment.

After spending time with the Del Castillo's, I reaffirmed that the simple, everyday rituals of purchasing fresh ingredients from local markets, preparing meals with purpose, and sharing food with loved ones were the cornerstones of happiness. The family's unpretentious approach to food and their sincere passion for creating and enjoying meals together left a lasting impression on me. This heartwarming experience fortified my belief that the true essence of gastronomy lies in the love, care, and joy shared through everyday culinary traditions.

GAZPACHO

Serves: 4-8
Prep Time: 15 minutes
Cook Time: 0 minutes

840 g fire roasted red peppers
450 g campari tomatoes
1 cucumber, peeled and quartered
1/2 small shallot, peeled and chopped
15 ml extra virgin olive oil
10-30 ml red wine vinegar
Salt and pepper, to taste

Directions

1. Combine all the ingredients together in a large bowl to use an immersion blender or place ingredients in a blender.
2. Blend until desired smoothness.
3. Serve immediately or refrigerate in sealed glass container.

Notes

Try serving small portions in shot glasses with an edible flower placed on top.

Here is a heartwarming review from a rather unlikely source: a vegetable averse pastor from the southern United States. Despite firmly believing that vegetables had no place on a burger, he absolutely loved my Gazpacho. He told me that if I ever needed any contractor work done on my house or yard, he would be more than happy to do it, as long as he was paid in Gazpacho. This story really highlights the transformative power of food and the love that goes into creating a dish like Gazpacho. It is a reminder that sometimes, a simple recipe can create lasting connections and bring people from all walks of life together.

GLOSSARY

Al dente: Cooked so that it is still firm when bitten, often used to describe the ideal texture for pasta and vegetables.

Aromatics: Vegetables and herbs that provide a strong, distinctive aroma and flavor to dishes, such as garlic, onion, and celery.

Au gratin: A dish topped with breadcrumbs and/or cheese, then baked or broiled until browned and crispy.

Blanch: To briefly immerse vegetables in boiling water, then transfer to cold water to stop the cooking process. This technique is often used to enhance color, texture, and flavor.

Bouquet garni: A bundle of herbs tied together or wrapped in cheesecloth, used to add flavor to soups, stews, and stocks.

Braise: A cooking method where vegetables are first seared at high heat, then simmered in a small amount of liquid, often in a covered pot or pan.

Brunoise: A French culinary term for a very fine dice, typically 2-3mm cubes, often used for vegetables such as carrots, celery, and onion.

Caramelize: To cook vegetables over low heat until their natural sugars brown and develop a deep, rich flavor.

Chiffonade: A cutting technique where leafy vegetables or herbs are rolled tightly and sliced into thin, ribbon-like strips.

Concassé: A term used to describe vegetables, often tomatoes, that have been peeled, seeded, and chopped.

Confit: A cooking method where vegetables are slowly cooked in oil or fat at a low temperature, preserving their texture and enhancing their flavor.

Coulis: A smooth, thick sauce made from puréed vegetables or fruit.

Crudités: A French term for raw vegetables, often served as an appetizer with dip or dressing.

Deglaze: To add liquid, such as wine or broth, to a hot pan to loosen the browned bits of vegetables, adding flavor to the dish.

Dice: To cut vegetables into small, uniform cubes.

Emulsion: A mixture of two liquids that do not naturally mix, such as oil and vinegar, often used for salad dressings and sauces.

En papillote: A cooking method where vegetables are enclosed in parchment paper or foil and baked, steaming in their own juices.

Espuma: A light, frothy foam made from puréed vegetables, often using a whipped cream dispenser.

Floret: A small, individual piece of a larger vegetable, such as broccoli or cauliflower.

Fond: The flavorful browned bits left in a pan after cooking vegetables, often used as a base for sauces.

Julienne: A cutting technique that creates thin, matchstick like strips of vegetables.

Macerate: To soak vegetables in an acidic liquid, like vinegar or lemon juice, to soften and enhance their flavor.

Mirepoix: A mixture of diced onions, carrots, and celery used as a flavor base for many dishes, particularly soups, stews, and sauces.

Noisette: A small, round piece of vegetable, often used to describe the size and shape of potatoes.

Parboil: To partially cook vegetables by boiling them briefly before using them in another cooking method.

Pickle: To preserve vegetables in a brine or vinegar solution, often with added flavorings such as herbs and spices.

Purée: To blend or mash vegetables into a smooth consistency, often used for soups, sauces, or baby food.

Rémoulade: A French sauce made from mayonnaise, mustard, capers, and chopped pickles or gherkins, often served with vegetables.

Roast: A cooking method that involves baking vegetables in an oven at a high temperature, usually with added fat, to develop a deep, rich flavor and crisp texture.

Roux: A mixture of equal parts fat (usually butter) and flour, used as a thickener for sauces, soups, and stews.

Sauté: A cooking method that involves quickly frying vegetables in a small amount of fat over high heat.

Slaw: A salad made from shredded or chopped vegetables dressed with a creamy or vinaigrette-based dressing.

Steam: A cooking method that involves placing vegetables over boiling water, allowing the steam to cook them gently and evenly.

Stir fry: A cooking method that involves quickly frying vegetables in a small amount of oil over high heat, typically in a wok.

Succotash: A dish made from cooked corn and lima beans, often with additional vegetables such as bell peppers, onions, or tomatoes.

Sweat: To cook vegetables gently in a small amount of fat, usually covered, so that they release their moisture and become tender without browning.

Tempura: A Japanese cooking method where vegetables are dipped in a light batter and deep-fried until crispy.

Terrine: A layered dish of vegetables, often combined with meat or seafood, that is cooked and served cold, typically in a loaf shape.

Umami: A savory taste, often associated with foods rich in glutamate, such as tomatoes, mushrooms, and soy sauce.

Vinaigrette: A dressing made from oil and vinegar, often seasoned with herbs, spices, and other flavorings.

Wilt: To cook leafy vegetables briefly, often in a small amount of fat, so that they soften and become tender.

Zest: The colorful outer layer of citrus fruit, often used as a flavoring or garnish in recipes.

CONVERSIONS

Standard Unit	Metric Unit
1 teaspoon	5 ml
1 tablespoon	15 ml
1 fluid ounce	30 ml
1 cup	240 ml
1 pint	473 ml
1 quart	946 ml
1 gallon	3.785 L
1 ounce	28.35 g
1 pound	453.59 g
1 inch	2.54 cm
1 foot	30.48 cm
1 yard	0.914 m
1 mile	1.609 km

Made in the USA
Monee, IL
15 August 2023

03e55917-53a4-4f75-98a4-501dbe182d1dR01